CULTIVATING WOMEN, CULTIVATING SCIENCE

❧ ❧ ❧ ❧ ❧ ❧ ❧ ❧ ❧ ❧ ❧ ❧

Flora's Daughters
and Botany in England
1760 to 1860

❧ ❧ ❧ ❧ ❧ ❧ ❧ ❧ ❧ ❧ ❧ ❧

CULTIVATING
WOMEN

CULTIVATING
SCIENCE

ANN B. SHTEIR

THE JOHNS HOPKINS UNIVERSITY PRESS
BALTIMORE AND LONDON

Published in cooperation with the Center for American Places, Harrisonburg, Virginia

© 1996 Ann B. Shteir
All rights reserved. Published 1996
Printed in the United States of America on acid-free paper
05 04 03 02 01 00 99 98 97 96   5 4 3 2 1

The Johns Hopkins University Press
2715 North Charles Street
Baltimore, Maryland 21218-4319
The Johns Hopkins Press Ltd., London

LIBRARY OF CONGRESS CATALOGING-IN-PUBLICATION DATA

Shteir, Ann B., 1941–
    Cultivating women, cultivating science : Flora's daughters and botany in England,
1760–1860 / Ann B. Shteir.
        p.   cm.
    Includes bibliographical references (p.   ) and index.
    ISBN 0-8018-5141-6 (hc : alk. paper)
    1. Botany—England—History—18th century.   2. Botany—England—History—
19th century.   3. Women in botany—England—Biography.   I. Title.
QK21.G7S57   1996
581'.082—dc20   95-12736

A catalog record for this book is available from the British Library.

Illustration, title page: Anne Pratt (1806–93), detail. (Courtesy of Hunt Institute for Botanical Documentation, Carnegie Mellon University, Pittsburgh)

To my mother and my daughter, and to the memory of my father

Contents

Acknowledgments

The research embodied in this book has been supported by grants from the Social Sciences and Humanities Research Council of Canada and from Atkinson College and the Faculty of Arts, York University. The award of an Atkinson Research Fellowship came at the right moment. It is a pleasure to acknowledge research assistants who helped me in this work, especially Katherine Binhammer, Philippa Schmiegelow, Yvette Weller, and Jessica Forman.

Over ten years ago the late Blanche Henrey, author of *British Botanical and Horticultural Literature before 1800* (1975), invited me to tea, showed me her collection of eighteenth-century books, and engaged with me in botanical conversation about women whose stories had not been told. In more recent years David E. Allen has offered many rounds of information and bibliographical tips, drawing on his unparalleled fund of knowledge about the history of natural history. I gratefully acknowledge conversations with Ray Desmond, Anne Secord, and Jim Secord. For substantive help on aspects of my research, I am indebted to Christopher Roper, Desmond King-Hele, Alan Bewell, Clarissa Campbell Orr, Dawn Bazely, Nancy Dengler, Marion Filipiuk, Harold B. Carter, Guy Holborne, and David Trehane.

Materials for this study were often elusive and called for archival digging, wide consultation, and good fortune. I express my gratitude to Interlibrary Loan at York University and to the staff of numerous research libraries, particularly Gina Douglas of the Linnean Society, London; John Symon, Wellcome Institute Library, London; Malcolm Beasley at the Botany Library, Natural History Museum, London; British Library; Friends' House Library, London, particularly Malcolm Thomas; David Doughan, Fawcett Library, London Guildhall University; Anita Karg and Charlotte Tancin at the Hunt Institute for Botanical Documentation at Carnegie-Mellon University; Dana Tenney, Margaret Maloney, and Jill Shefrin at the Osborne Collection of Early Children's Books, Toronto;

the New York Public Library; and the Library of Royal Botanic Garden, Kew. I consulted materials in the Record Offices of Leicestershire, Lancashire, Devonshire, and Derbyshire and acknowledge assistance from staff members.

I have reported on this work at many conferences and research seminars, and I want to note in particular meetings of the American Society for Eighteenth-Century Studies; Berkshire Conference of Women Historians; Cabinet of Natural History, Cambridge University; History of Science Society; Society for Literature and Science; Women's Studies Group 1500–1830, London; and Graduate Women's Studies Programme Seminar Series, York University. Very special thanks go to students in my graduate seminars at York University on "Women and Eighteenth-Century Writing."

Sten Kjellberg, Frederick Sweet, and Christina Erneling translated materials for me from Swedish and Latin. For technical assistance I thank the Photographic Division of the Wellcome Institute for the History of Medicine, London, and Hazel O'Loughlin-Vidal and John Dawson of York University.

Some material has appeared in different form in essays published earlier. I wish to acknowledge the following: "Botanical Dialogues: Maria Jacson and Women's Popular Science Writing in England," *Eighteenth-Century Studies* 23 (1990):301–17; "Botany in the Breakfast-Room: Women and Early Nineteenth Century British Plant Study," in *Uneasy Careers and Intimate Lives: Women in Science, 1789–1979*, ed. Pnina G. Abir-Am and Dorinda Outram, 31–43 (New Brunswick: Rutgers University Press, 1987); "Flora Feministica: Reflections on the Culture of Botany," *Lumen* 12 (1993):167–76; "Linnaeus's Daughters: Women and British Botany," in *Women and the Structure of Society: Selected Research from the Fifth Berkshire Conference on the History of Women*, ed. Barbara J. Harris and Joann K. McNamara, 67–73 (Durham, N.C.: Duke University Press, 1984); and "Priscilla Wakefield's Natural History Books," in *From Linnaeus to Darwin: Commentaries on the History of Biology and Geology*, 29–36 (London: Society for the Study of Natural History, 1985).

David E. Allen, Donna T. Andrew, Frieda Forman, Joan Gibson, Bernard Lightman, and Peter F. Stevens read chapters. Barbara T. Gates, Cynthia Zimmerman, and Grif Cunningham each read the entire manuscript at different stages. I cannot thank them all enough. My gratitude goes to the Readers for the Johns Hopkins University Press for their suggestions, and to Alice M. Bennett for her work on the manuscript.

For the friendship of scholarship and for their expressions of enthusiasm and confidence in the project, I warmly thank Janet Browne, Harriet Ritvo, Mitzi Myers, Judith Phillips Stanton, Ruth Perry, Madelyn Gutwirth, Estelle Cohen, Isobel Grundy, Jan Fergus, and Gina Feldberg. For patience and good cheer, my appreciation goes to George F. Thompson. I am also deeply grateful to others who kept me company during the long gestation of this book: Sheila Malovany Chevallier, Etta Siegel, Marilyn Shteir, Joel Martin, Bronwen Cunningham, and Neill Cunningham. Grif Cunningham has given me the treasured gifts of companionship and understanding. Our daughter entered into the spirit of my project with grace and enthusiasm as our own feminist Flora.

CULTIVATING WOMEN, CULTIVATING SCIENCE

Botanical Conversations

On a cold early spring day in England during the late eighteenth century, two women walk together in a garden and stop to admire some blossoms just visible above the snow. One woman, a new student of botany, uses technical language to describe the flowers. Her teacherly companion, Flora, welcomes her enthusiasm for studying plants. *The New Lady's Magazine, or Polite and Entertaining Companion for the Fair Sex* carried a "Botanical Conversation" in 1786 that read as follows:

> INGEANA: How charming these snow-drops still look; notwithstanding the late frost, and the depth of the snow, with whose whiteness they seem to vie.
>
> FLORA: The snow, my dear, has preserved both their beauty and life; otherwise they must have fallen a sacrifice to the severity of the weather.
>
> INGEANA: What elegant simplicity and innocence in this flower! It belongs, I believe, to the sixth class of the Linnean system, called Hexandria, and by our botanical society Six Males, and to the first order of that class: but it seems to me to be two flowers, a less within a greater.
>
> FLORA: The whole is but one flower; this part, which you suppose to be a lesser flower, is called by our ingenious translators of the immortal Linneus, the nectary; and indeed emphatically; for if the bee were now stirring, you would see him drink his honey out of it.
>
> INGEANA: I have often admired the green streaks on each of these shorter petals of the nectary; eight in number; which through a microscope swell on the eye like a piece of beautiful fluted work: but we are called.[1]

The remarks by Flora and Ingeana, blending aesthetic commentary and scientific detail, refer to the influential Swedish botanist Carl Linnaeus, whose ideas about plant classification and nomenclature helped popularize botany in England in the eighteenth century. Their brief conversation is interrupted at a point when it might become a more technical elaboration of Ingeana's findings from her work with a microscope. The two women are called away from their botanical moment, perhaps to more conventional domestic duties.

Flora and Ingeana are important figures in the history of women, writing, and science culture. They appeared in a magazine catering to women readers in the middle ranks of society at a time when fashion, commerce, and social values put science on the list of culturally approved activities. During the European Enlightenment, science learning was part of general and polite culture. Women were cultivated as consumers of scientific knowledge, and many books and periodicals introduced "the fair sex" to sciences of the day.[2] Promoters of science for women recommended astronomy, physics, mathematics, chemistry, and natural history as activities for moral and spiritual improvement. They represented science as an antidote to frivolity and an alternative to the dangers of the card table. Women who studied science, they maintained, would become better conversationalists and more successful mothers. After 1760 botany became a particularly fashionable pursuit in England, and conversations about plants and flowers took place both in print and in everyday life.

Then as now, plants and flowers were used, valued, and studied in social and cultural domains as diverse as horticulture, aesthetics, medicine, taxonomy, recreation, geography, religion, and commerce. Then as now, they carried different meanings for various communities of practitioners.[3] During the eighteenth and nineteenth centuries botany bridged aesthetic, utilitarian, and intellectual approaches to nature. The impetus to study plants came, for example, from interest in plant physiology, evolution, and the classification and order of nature, or it had spiritual and social motivations. For some the study of botany represented intellectuality, for others social cachet, social reform, spiritual satisfaction, or economic advantage. For some botany was a theoretical area of inquiry quite different from the applied knowledge of horticulture.

In one eighteenth-century representation of botany, Flora appears as a flower-bedecked goddess. She joins the goddess of agriculture and the gods of healing and love in the portrait *Aesculapius, Flora, Ceres, and Cupid Honouring the Bust of Linnaeus*, an illustration from Robert Thornton's *New Illustration of the Sexual System of Carolus von Linnaeus* (1799). The teacherly Flora of the "Botanical Conversation" bears a name saturated with eighteenth-century cultural assumptions about women. Her

Aesculapius, Flora, Ceres, and Cupid Honouring the Bust of Linnaeus, from Robert Thornton's *New Illustration of the Sexual System of Carolus von Linnaeus* (1799). (By permission of The British Library, shelfmark 10.Tab.40)

name also resonates with traditional associations from myth and literature that link flowers and gardens with women and nature and with femininity, modesty, and innocence. During the eighteenth and nineteenth centuries, cultural linkages like these helped smooth the path for women into botanical work of many kinds. Encouraged by parents, teachers, and social commentators and pursuing their own interests, Flora's English daughters were botanically active. They read botany books, attended public lectures about plants, corresponded with naturalists, collected native ferns, mosses, and marine plants, drew plants, developed herbaria for further study, and used microscopes. Of special interest in my study, Flora's English daughters also wrote about botany, and their books, essays, and poems are a rich cultural resource for chronicling the experiences of young girls, women, and mothers in the science culture of their day.

The year 1760 marks the time when the spread of the Linnaean system, a classificatory scheme for grouping and naming plants, contributed to cultivating interest in botany in England. The ease and simplicity of this system recommended botanical study to men, women, and children. During the period 1760–1830, botany was constructed as both a fashionable and an "improving" pursuit in line with social and cultural values. Chapter 1 charts botany as a cultural feature in eighteenth-century England and shows how it was promoted by books for different levels of readers. Since the Linnaean system was based on the "male" and "female" reproductive parts of flowers, it put the topic of sexuality before those who studied plants. Cultural tensions about women, gender, sexuality, and politics clustered around this issue. Women's botanical activities were configured by gendered beliefs about women as students and readers, teachers and writers.

Chapter 2 introduces women who were botanically active during the eighteenth century. In earlier times women had worked with plants in the herbal tradition of home-based medical practices, but in the eighteenth century that familiarity was supplanted by a new polite culture of botanical art and fashion. In the history of botanical culture, daughters worked alongside their fathers, and girls' botanical interests developed within botanical families. Some women studied Latin and corresponded with botanists, including Linnaeus himself. Although women's work in botany was shaped by gendered beliefs about limits to female learnedness, women also shaped botanical knowledge to their own ends. Thus Anna Garthwaite, a midcentury fabric designer, brought botanical naturalism into her work and put plants with roots on the damask gowns of the aristocracy.

Botany became part of the gender economy for women in England, and they could make it work for them. During 1790–1830 women were particularly visible as writers of botany books. Cultural discourse and social norms gave new prominence to their roles as mothers and educators. A new maternal ideology in the eighteenth century lent authority to women in scientific education and popular science writing. Writers using the "familiar format" of letters and conversations featured women teaching children at home, thereby authorizing them by providing enabling models. Botany as a popular subject for informal education also attracted women who wanted to earn money by writing for the juvenile market, women, and general readers. Chapters 3 and 4 introduce women who wrote about botany in genres such as verse, juvenile natural history books, novels, and introductory books for family-based education. Profiles of writers such as Charlotte Smith and Priscilla Wakefield illustrate the material conditions of women's authorship within science culture.

Chapter 5 focuses on three women from the turn of the nineteenth century who made botanical topics central to their writing lives. The stories of Maria Jacson, Agnes Ibbetson, and Elizabeth Kent all intersect with science culture and literary culture, extending from the Enlightenment decades across the Linnaean years and into romanticism. These three women stand at the heart of this study of Flora's English daughters: working within changing parameters of beliefs about women, gender, and science, each struggled to make her contribution, and each also made botany a resource for herself. Each can be said to have had a "career" as a botanical writer.

By 1830 botany began to have a modern profile as a science, and women's status as writers and cultural contributors began to be considered problematic. Chapter 6 sketches these changes in botanical culture across the middle of the nineteenth century. John Lindley, the first professor of botany at the University of London, is a central figure in this story, for he wanted to modernize and defeminize botany. His determination to forge distinctions between polite botany—what he called "amusement for ladies"—and botanical science—what he called "an occupation for the serious thoughts of man"[4]—inserts a highly charged element of gender into the professionalizing of science.

During 1830–60, when ideas about science were in flux, many women participated eagerly in botanical activities at home and abroad. Chapter 7 introduces a series of botanizing individuals, including young fern collectors, prolific watercolorists, a wax-flower maker and teacher, and avid marine botanists, as well as colonial wives who were commissioned to send home specimens for imperial projects. In chapter 8 discussion of writings from the early Victorian years delineates botany as a continuing textual resource for women. Women continued to write for children, women, and general readers during the time when generalist and specialist approaches to science were diverging. Well-known figures such as Anne Pratt and Jane Loudon illustrate this rich mid-nineteenth-century history, as do figures whose stories are less familiar to us now. Their books show that narrative forms for introductory texts were changing as family-based conversations were replaced by the impersonal accounts familiar to us as the standardized format of modern science books.

In the "Botanical Conversation" I began with, Flora and Ingeana discuss not fashion or families but science—botanical nomenclature and taxonomy at that. Their presence has been restored to visibility within the history of science, literary history, the history of education, and the history of gender ideology in modern European culture. Conventional histories of botany, in line with traditional disciplinary assumptions, report on heroic (male) individuals and scientific advances; it is a historiographic style that excludes the kinds of botanical work women did and could do in earlier periods.[5] By contrast, female botanical and scientific conversationalists figure in social, cultural, and literary histories of science and natural history.[6] Within the fields of women's studies and the history of science, researchers who turn the lens back to women in preprofessional science have explored science as part of the lives and the work of girls and women in earlier periods. Developments in the history of the family, the institutionalization of science, and gender ideology are among the features of their stories.[7] Studies of the rhetoric of science, bringing a text-based focus to science culture, have shown how literary practices embody social structure and themselves shape knowledge. It has become clear, for example, that genres of science writing such as the dialogue have their own gendered histories and literary, sociohistorical, and political frameworks.[8] There has been a recent call for a feminist historicism attentive to the diversity of women's writing in earlier periods,[9] and new attention is being directed to popular science writing and periodical articles as resources for studying science culture. Current studies also have begun to explore how women tell the story of science.[10]

In the social and literary history of botany and natural history, Flora is not only a culturally resonant name, a visual symbol and emblem, but also a kind of writing. A flora is a genre in the natural history field tradition of observation and classification; William Curtis's *Flora Londinensis* (1775–98), for example, lists the plants of the area and organizes them according to a taxonomic system. My study of Flora's daughters in England is a flora too, a handbook about women who were part of the culture of botany during the formative years of modern science culture. While my own emphasis is on women writers within the culture of botany, the textual and biographical material also invites attention to other topics, such as women and scientific language, consumerism, readership, religion, gardening, and the herbal tradition and domestic medicine.

This book features botany as a scientific and literary undertaking for actual women and focuses on botany as an occupation of choice, part of the parcel of normative activities for girls and women. I highlight actual women rather than the discursive Woman because I believe that research in women's writing and women's history needs to stay (as it were) grounded. In the richness of contemporary theorizing, we have honed and reconfigured methods of interpretation and brought new analytic tools to our research. But as we dig deeply into literary and cultural history and perform sophisticated soil probes in our intellectual and political terrain, as we turn over the older categories, letting them compost into new interpretive riches, we should not forget the women and their individual stories.

ONE

Spreading Botanical Knowledge throughout the Land, 1760–1830

[Botany is] as healthful as it is innocent . . . [it] beguiles the tediousness of the road, . . . furnishes amusement at every footstep of the solitary walk, and, above all, . . . it leads to pleasing reflections on the beauty, wisdom, and the power of the great CREATOR.

> William Withering, *A Botanical Arrangement of All the Vegetables Naturally Growing in Great Britain*, 1776

There are few studies more cultivated at present by persons of taste, than Botany; and certainly, of all those not immediately conducive to the wants of society and the necessities of life, none can be more deserving of regard. Whether we consider the effect of Botany as enlarging the sphere of knowledge, or as conducive to health and innocent amusement, it ought to rank very high in the scale of elegant acquirements.

> William Mavor, *The Lady's and Gentleman's Botanical Pocket Book*, 1800

In the opening years of the nineteenth century the entrepreneurial artist and printer James Sowerby issued a card game meant to teach how to name and classify plants. *Botanical Pastimes . . . Calculated to Facilitate the Study of the Elements of Botany* (ca. 1810) consists of cards with engravings of parts of plants along with botanical questions and answers. As a "familiar introduction to the Science," it offers "a series of Questions and exemplifications, calculated to render the first more uninteresting part of the study an agreeable Amusement."[1] Such a publishing venture assumes a commercial market for popular science, notably for pastimes involving plants. Material culture and changing sensibilities had brought the study of nature to prominence during the eighteenth century. England was inundated with new introductions of plants from exotic parts of the world, and attention turned to native flora as well. Botanic gardens were established, notably at Kew in 1759. Men, women, and children across the middle ranks of society took up botanical books and puzzles that gave them opportunities for a culturally sanctioned mixture of amusement and instruction. By the century's end botany was part of culture and commerce, just as demonstrations of experiments in astronomy and natural philosophy had been earlier in the century.

As part of Enlightenment science, botany benefited from the cultural enthusiasm for scientific activities during the eighteenth century. In England as on the Continent, inquirers grounded in the New Science sought new kinds of information, developed new approaches to knowledge, and put their ideas into practice in new ways (promoting Newtonian natural philosophy, for example, for commercial and entrepreneurial reasons). Science culture during the European Enlightenment concerned itself with the creation of new scientific ideas and the dissemination of ideas for social and individual ends. Natural history and science were on many different social, cultural, and political agendas. Scientific societies developed and spread, advancing scientific practices, communicating scientific findings, and generally promoting science. Enlightenment thinkers and writers wanted to demystify knowledge and bring formerly arcane material into the public arena for broader public scrutiny, and science education therefore featured in the mandates of Enlightenment science culture.[2]

Botany manifested the same classifying and ordering impulse that was

Three cards from a set titled *Botanical Pastimes . . . Calculated to Facilitate the Study of the Elements of Botany*, ca. 1810. (By permission of the Trustees of The Natural History Museum, London)

evident in areas such as physics and linguistics, where scholars during the seventeenth and eighteenth centuries searched for universal laws. Botanists believed there were discoverable patterns in nature and that a natural and a rational classification of plants was discernible. But how is such a natural system of relationships among plants to be described? Does one describe plants by reference to size, habitat, uses, or variations in color or taste, or in terms of features during the growth cycle, such as parts of the flower or the seed? Botanists put together organizational schemes to represent relationships within the amplitude and variety of the vegetable kingdom and to specify the distinguishing features, or characters, of different plant groupings. John Ray, for example, propounding a natural system in taxonomy, believed that plant classification should be based on several morphological features of a plant; his French contemporary, Joseph Pitton de Tournefort, grouped plants according to the structure of one single part of the flower. Tournefort's system of plant classification, widely used as a practical tool for grouping plants during the eighteenth century, served heuristically for teaching botany, and aided the spread of

Carl Linnaeus. This line engraving, made in 1788, is based on a portrait by Roslin that was copied for Sir Joseph Banks. (Courtesy of the Linnean Society, London)

botanical knowledge from the schooled and the experts to ever increasing numbers of botanical enthusiasts.

In England, the artificial system for classifying plants that was promulgated by the Swedish naturalist Carl Linnaeus gave botany a special boost, and the Linnaean system played a central part in making botany accessible to different groups and levels of enthusiasts. Linnaeus (also known as Carl von Linné) based his taxonomy on one diagnostic feature of a flower—the reproductive parts. Linnaeus's sexual system, as it was called, assigned taxonomic centrality to the part the flower plays in plant reproduction. He divided the plant kingdom into classes and orders by singling out the male and female reproductive parts of plants. He categorized plants by counting, first, the (male) stamens and their proportions and then the

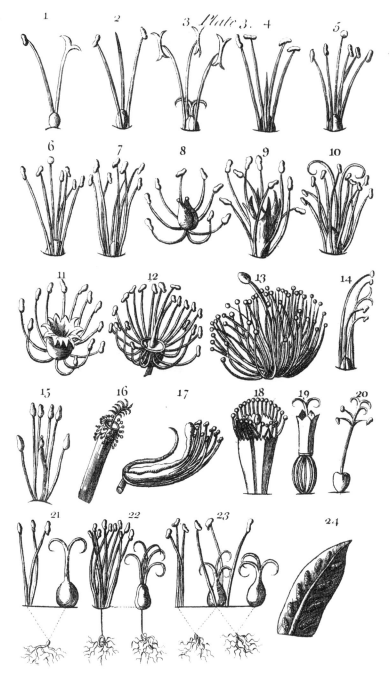

James Lee, *Introduction to Botany: Extracted from the Works of Dr. Linnaeus* (1760), plate 3, "Linnaean Classes."

(female) pistils. Based on the number of stamens in a flower, Linnaeus divided the vegetable kingdom into twenty-three classes, plus one class for nonflowering plants. Linnaeus coined names for classes based on the root *andria* ("male"); thus flowers with one stamen became class 1, Monandria; those with two became class 2, Diandria; and so on. He then divided classes into orders based on the characters of the female part. He chose the root *gynia* ("female") for that designation. Thus, flowers with one pistil became Monogynia.[3] The snowdrop of the 1786 "Botanical Conversation" with its six stamens and one style therefore belongs to the Linnaean class Hexandria and the order Monogynia.

As a theory about forms of order in nature, the Linnaean system, like any explanatory system, had many social meanings and was used for several different and often contrasting purposes. An occasional poem from the 1780s, for example, represented Linnaean botany as a panacea for social disorder. "The Backwardness of the Spring Accounted For" depicts Jupiter descending to earth on May Day to look for signs of a burgeoning new season.[4] Jupiter finds instead that Flora's kingdom is in disarray, in a "confusion of Manners and Morals." Flowers are acting out of turn, not respecting their elders and betters; the "Vagabond Fungus" is treading "on the toes of his highness the Oak." Why, Jupiter asks Flora, is this so? Flora sadly reports that her reputation is not honored in Great Britain, and that

> . . . an ignorant treatment of her and her System
> Encouraged her subjects she said to resist them.
> Vegetation of course was o'errun with disorder
> From the wood and the wall to the bank and the border.

According to Flora's report, her kingdom is in disorder because her subjects fraternize inappropriately and consort with radically divergent flowers. They are not honoring their proper places in her kingdom. But help is at hand and will come from social legislators who will instruct her subjects in the Linnaean system. Some scholars, Flora explains, are busy "compiling my Classification":

> These great legislators will shortly prescribe
> The laws, rules and habits of every tribe;
> Thus their manners no longer each other will shock:
> What is wrong in a rose may be right in a dock.

Speaking beyond Jupiter to her kingdom at large, Flora makes a crowning declaration of faith and comfort:

Rejoice then, my children, the hour is at hand
When botanical knowledge shall govern the land.

The whimsy of the poem should not mask its implicit argument about
the value of the Linnaean system. Flora exalts and legitimates decorum,
hierarchy, and traditional authority. In the poem's social vision, plants
know their place.

Linnaeus was one of a number of botanists during the late seventeenth
and early eighteenth centuries who explained the reproduction of plants
in terms of sexual reproduction.[5] Drawing on experiments about pol-
lination, hybridization, and the development of seeds, he argued from
analogy between plant functions and animal functions and referred to
earlier discussions about "male" and "female" plants, the sexual nature
of pollen, and "fertility" and "barrenness" in the plant kingdom. Linnaeus was awarded a prize in September 1760 in a competition offered
by the Imperial Academy at Saint Petersburg for a discussion about plant
sexuality. There he argued by analogy to male and female reproductive
organs, female "genital fluid," and "male pollen" and took the position
that stamens are the male part and pistils the female part of a flower.[6]

Linnaeus was generally measured and earnest in presenting his ideas,
but some of his narratives offer a spirited and imaginative rendering of
vegetable sexual relations suffused with eighteenth-century sexual politics. In one of his earliest writings, the *Praeludia Sponsaliorum Plantarum*
(1729), Linnaeus gave anthropomorphized accounts of brides and bridegrooms, of marriage and conjugal coupling, of "clandestine marriages"
and "eunuchs" in the plant kingdom. The sexuality that infuses all parts
of nature's realm, so his implicit argument goes, also shapes the conduct
of male and female, assigning each a preordained part to play in continuing the cycle of life. Linnaeus presents as "natural" what we should read as
his discourse about sexuality, a discourse he shares with other proponents
of the theory that the reproductive system of plants is directly analogous
to human sexuality. Linnaeus's assumptions about sexual functions in
plant reproduction reflect traditional gender concepts. Thus he assigned a
higher ranking to the class, a unit based on the stamens (the male part),
and a subsidiary ranking to the order, based on the pistil (the female part).
He also represented the male part in plant reproduction as active and the
female part as receptive. At the least, we can say that he naturalized sex
and gender ideologies of his day, folding into his botanical system ideas
of his own that also replicated beliefs of his day. The Linnaean taxonomy,

in other words, mirrored Linnaeus's own beliefs as well as the world of social relations of early and mid-eighteenth-century Europe.[7] Linnaean botany also can be read as constructing difference. The Linnaean system of plant sexuality was promulgated during the period when sexual difference became a reiterated theme in medicine and other areas of natural knowledge.[8]

Linnaean botany expresses a contemporary obsession with sex and gender difference. Tensions about defining gender boundaries are rife across eighteenth-century culture, suggesting gender anxieties. Ambiguities about sex and gender would challenge any taxonomist who craved clarity and simplicity in categories. Linnaeus's sexual system reads as a conservative gender construction. It embodies clear and naturalized sexual differences and distinct gender boundaries; it asserts the biological incommensurability of the categories "male" and "female." His highly naturalized and gendered theory about male/female difference in plant reproduction can therefore be read as illustrating a larger moment of reaction to cultural fears about blurred distinctions in sex and gender, and to gender ambiguity and shifting sex roles.[9]

Linnaeus's ripe and pictorial language of plant sexuality gave his system a distinctive brio for some botanists while arousing the displeasure and disapproval of others. Objections to the "sexualists" came from those whose experiments showed different results and from those who were skeptical about argument by analogy. Charles Alston, professor of medicine and botany in Edinburgh, one of the more vociferous opponents of the Linnaean sexual system, believed that Linnaeus generalized too much from some of his experiments and that explanations other than plant sexuality could account equally well for the lack of fruit on a tree. Alston, writing to stem the Linnaean tide, argued that pollen is "rather excrementitious and noxious, than useful to the nourishment or foecundity of the seed." He maintained that the sexual system "has given occasion to an intire deformation of botany, and to the introduction of an infinite number of new names, and perplexing tho' childish terms, whereby this most useful science is like to become not only vastly more difficult, but even ridiculous." Alston also objected to Linnaeus's way of sexualizing plants. Linnaeus's observations on a particular flower were, he wrote, "too smutty for British ears."[10] Alston, the conservative botanist, is responding to a theory about plant sexuality that Linnaeus himself sexualized and eroticized.

In England, various pioneering books introduced readers to Linnaeus's ideas about classifying and naming plants. Some were expositions of the Linnaean system, others applied his classifications to indigenous and regional plants or to plants of more remote areas. Adherents of the system had begun to translate Linnaeus's writings into English in the late 1750s. Benjamin Stillingfleet, a congenial man of letters associated with Bluestocking circles, who became a committed Linnaean in the 1750s, compiled *Miscellaneous Tracts relating to Natural History, Husbandry, and Physick* (1759), a translation of six dissertations done under Linnaeus's close supervision, and added his own essay in praise of Linnaeus and his botanical system. In 1760 James Lee issued his *Introduction to Botany*, a pioneering book about the Linnaean system. It consists of translated extracts from Linnaeus's writings, especially *Philosophia Botanica*, in which Linnaeus explained his system of classification and the technical terms he used for parts of the flower. Lee, a trendsetting nurseryman and proprietor of the well-known Vineyard Nursery in Hammersmith, London, was responsible for introducing many exotic plants into cultivation in English gardens, among them the fuchsia. He wrote his book "to assist such as may be desirous of studying [Linnaeus's] Method and Improvements," and he explained in the preface that "though the Study of Botany is of late Years become a very general Amusement in this Country, there has yet appeared no Work, in our own Language, that professedly treats of the Elements of that Science." Additionally, available expositions of the Linnaean system were "too costly to fall into many Hands." An appendix lists over two thousand English plant names and their Linnaean class, order, and genus. James Lee's *Introduction to Botany* was an early Linnaean best-seller and remained a standard introductory work for fifty years.

During the 1760s Linnaean ideas took root and spread in England. Looking back on this period Thomas Martyn, professor of botany at Cambridge and a Linnaean disciple, reflected in a letter to Sir James E. Smith, founder of the Linnean Society, on "how dead botany was in England in the middle of the last century. . . . Linnaeus's . . . system can hardly be said to have been *publicly* known among us till about the year 1762, when Dr. [John] Hope taught it at Edinburgh, and I at Cambridge, and Hudson published his *Flora [Anglica]*."[11] The taste for botany was cultivated in print and on the ground. William Curtis organized field trips around London during the 1770s for gentlemen who "wish to acquire a knowledge of the Plants growing wild about town, or of Linnaeus's System of Botany." Curtis was employed by the Worshipful Society of Apothecaries of London in the long-established post of demonstrator of plants at their teaching establishment, the Chelsea Physic Garden, and his duties included organizing "herborizings" for the apprentice apothecaries to enable them to recognize medicinal herbs in the wild. He went on to establish a private botanical garden of his own; subscribers to the garden were entitled to daily access to its plant collection and library.[12]

Publishers tried to capitalize on the taste and fashion for botany. Charles Bryant incorporated Linnaean names into his *Flora Diaetetica* (1783), a manual "portable in the pocket" about edible plants. He listed roots, berries, and nuts under their Linnaean binomials, to help "Gentlemen and Ladies not at all acquainted with Botany, . . . amuse themselves in their gardens, and examine the greatest part of their vegetables scientifically, without the fatigue of regularly studying the Science" (ix). In 1787 William Curtis founded the *Botanical Magazine*, an illustrated journal intended, so the title page proclaimed, for "such ladies, gentlemen and gardeners, as wish to become scientifically acquainted with the plants they cultivate." This magazine, renamed in later years *Curtis's Botanical Magazine*, appeared on the first day of the month and featured hand-colored engravings of plants with Linnaean names and information about cultivation. An immensely popular and frequently reprinted magazine ("upwards of 5,000 copies of each of the early numbers were sold"), it "gave respectability and authority to Linnaeus's ideas in England" by adopting Linnaean classification.[13]

Increasingly after 1760 botany books were written, and translated, with female readers in mind. Jean-Jacques Rousseau had written about botany for women in a work that became known as his *Lettres élémentaires sur la botanique* (1771–73). He drafted eight letters instructing a young mother about botany and how to study it, so that she could teach her own daughter in turn.[14] These letters have particular importance for British botanical culture because the botanist Thomas Martyn translated them into English so that "they might be of use to such of my fair countrywomen and unlearned countrymen as wished to amuse themselves with natural history," adding notes and twenty-four letters of his own "Fully Explaining the System of Linnaeus." The Rousseau-Martyn *Letters on the Elements of Botany* (1785) went through eight editions over the next thirty years as a very influential piece of popular botany writing. Retaining Rousseau's epistolary format and addressing letters to "my dear cousin," Martyn gave a Linnaean slant to the teachings that differed from Rousseau's own. Martyn explains that Rousseau laid a foundation for studying botany by providing some facts and a philosophy of nature. Martyn aimed to raise

William Mavor, *The Lady's and Gentleman's Botanical Pocket Book* (1800).

the "superstructure" by teaching his female pupil to identify plants herself. Rousseau had been antipathetic to systemizing and to any focus on names of plants; but how, Martyn asks, "is it possible [for every reader of common sense] to attach one's self to the study of plants, and at the same time to reject that of the nomenclature?" Martyn proposed the Linnaean system: "Not the French one by Tournefort, which is very beautiful, and has great merit; but the Swedish one by Linnaeus. I prefer this, because it is most complete, and most in fashion."[15] Joining the train of Linnaean botany for women, G. Gregory's *The Economy of Nature* (1796) was intended for "all who are curious about a general survey of nature, [and] young students of medicine, [but especially] the more enlightened class of female readers" (preface). Charles Abbot designed his *Flora Bedfordiensis* (1798), a Flora of Bedfordshire that lists and describes wild plants of that county along Linnaean lines, "for the amusement and instruction of 'the fair daughters of Albion.'" William Mavor's *The Lady's and Gentleman's Botanical Pocket Book* (1800), a "novel attempt to render botany still more fashionable" with the public, was organized as a research record for entering botanical findings in their appropriate Linnaean categories, with space for filling in the names of specimens one found.

Soon after Linnaean ideas were disseminated to general audiences in England, the conjunction of women readers and botany books created difficulties that were textual/sexual. The theories about plant sexuality central to Linnaeus's ideas about plant classification made botany a highly charged area for writers, teachers, parents, translators, and publishers who wanted to promote the study of botany.

During the late 1770s, motivated by an Enlightenment commitment to the usefulness of knowledge, the physician-botanist William Withering set out to write a popular manual of botany. His *Botanical Arrangement of All the Vegetables Naturally Growing in Great Britain* (1776), a handbook of plant classification and an introduction to the Linnaean system, was particularly well known and went through many editions; the third edition of 1796, much enlarged and revised, was considered authoritative. "Withering's Botany," as the book came to be known, remained "the standard text on British plants for at least a generation."[16] Readership ranged from expert Linnaeans to keen amateurs and children. Dorothy Wordsworth wrote in her *Grasmere Journal* on May 16, 1800: "Oh! that we had a book of botany." She and her brother William acquired a copy of the third edition of Withering's book later that year.[17] Twenty years later a character in a popular children's book from 1823 complained about systematic botanical study by exclaiming, "I do not like your Withering at all, Emmeline!" Her botanically more engaged companion replies: "Stop, stop, my dear! You must not dislike poor Withering without reason, for he is a great favourite of mine, and his botanical arrangement of British plants is too valuable to be despised."[18]

The history of Withering's botany book crystallizes problematic features of Linnaean botany, for him in his day and for us in ours. William Withering used the Linnaean system in his *Botanical Arrangement*, and his text became the hub of a controversy about two issues involved in adapting the botanical systematics of Linnaeus for a wide and general audience. The first concerned how to translate Linnaean terminology from Latin

William Withering. (Courtesy of Hunt Institute for Botanical Documentation, Carnegie Mellon University, Pittsburgh)

into English, and the second concerned how to present Linnaean plant sexuality.

William Withering belonged to the Lunar Society of Birmingham, an informal group of gentlemen amateurs and friends in the Midlands who met regularly from 1775 into the early 1790s to discuss projects at the nexus of science and technology. Members such as Erasmus Darwin, Josiah Wedgwood, and Matthew Boulton took the wide-ranging intellectual findings of eighteenth-century science and applied them in fields such as mechanics, mineralogy, and botany.[19] On Linnaean language and the foundational features of the Linnaean system, Withering and Erasmus Darwin held divergent views, however. Withering believed that English terminology would make the Linnaean system easier for a general audience to understand. Accessibility being his guiding principle, he translated Linnaeus's Latin terms into English. In the first edition of his book he translated "stamen" and "pistil" as "chives" and "pointals,"

and "calyx" as "impalement." He also translated Linnaeus's Latin coin-
ages for leaf shapes into English; a leaf with an uneven and wavy edge,
which Linnaeus termed "repandus," became "scalloped" in Withering's
account.

Withering's book gave various communities of readers access to Linnaean botany, but his progressive motivations about access to knowledge
were more than counterbalanced by conservative notions about one part
of his intended readership. He wrote with women readers in mind. A
sweet letter to his daughter in 1791 reads: "I have been writing for you,
and many others, to whom I hope that [my botany book] will prove a
source of useful information and elegant rational amusement."[20] But how
then to deal with sexual material in a book whose readership includes not
just women in general, but also one's daughter? How explicitly and color-
fully should he describe the Linnaean sexual system? Should one glory
in raunchy explications of polygamous cross-fertilization in the vegetable
kingdom? Is discreet circumlocution called for? Withering's own gender
ideology was that women were (or should be) modest and chaste and that
it would be inappropriate to parade plant sexuality in front of them. As a
result, he translated terms and titles in the first edition of his botany book
in 1776 so as to exclude sexual reference. One cannot tell from his terminology that male and female sexual parts are the key to plant classification
for Linnaeus and the Linnaeans. Withering is quite direct about why he
is bowdlerizing, declaring in the preface to the first edition that "from an
apprehension that botany in an English dress would become a favourite
amusement with the ladies, . . . it was thought proper to drop the sexual
distinctions in the titles to the Classes and Orders."

Erasmus Darwin adamantly disagreed with Withering's approach to
Linnaean language and contended that one learned botany best by reading literal translations of Linnaeus rather than texts that anglicized his
Latin terms.[21] In 1783, seven years after the appearance of William Withering's book, the Botanical Society of Lichfield issued an English translation
of Linnaeus's *Systema Vegetabilium*, a catalog of over fourteen hundred
plants divided into the twenty-four classes of Linnaeus's system, under the
title *A System of Vegetables*. Their intention, the translators explain, was to
"English" Linnaeus—that is, to give "a literal and accurate translation."
They locate their work in relation to other Linnaean translations and,
in particular, set it in opposition to Withering's *Botanical Arrangement*,
which they contend has "rendered many parts of [Linnaeus's] work unintelligible to the latin Botanist; equally difficult to the English scholar; and

[has] loaded the science with an addition of new words." Their argument is that Withering, in seeking to write a popular manual, translated terms into English that are better left in Latin. They also objected to Withering's omission of the sexual distinctions (preface).[22] The preface to the Lichfield translation of Linnaeus is a credo about accuracy in translation, directed to those who know Latin but want to have conversations about Linnaeus in English. The translators presented their task as trying to naturalize into English usage the Latin and Latin-type terminology Linnaeus developed. Thus they retained the Linnaean Latin term "calyx" for the leaves that surround the flower bud rather than following Withering's translation of "calyx" as "impalement." They acknowledged technical difficulties in honoring the language that Linnaeus formed, particularly in translating compound terms. (Linnaeus constructed compounds to describe shapes of leaves and flowers—for example, "oblonged-egg" and "axe-form," which he then combined further in ways that made for precision in Latin but for awkwardness in English.) Nonetheless, their aim was to "equal the conciseness and precision of the original."

The Botanical Society of Lichfield's *System of Vegetables* (which was followed by a translation of Linnaeus's *Genera Plantarum* under the title *The Families of Plants*) shaped a language for English botany that built bridges between those who knew Linnaeus only in Latin and those who had little or no Latin and were trying to learn botany through English. In their project to English Linnaeus and shape a botanic language, they drew on Samuel Johnson himself for advice, and thanked him in the preface for his assistance.

The controversy between Darwin and Withering concerned not only the technical language of botany but also the issue of plant sexuality, the central credo of Linnaean systematics. By anglicizing Linnaean terms, Withering deliberately masked the sexual character of the taxonomy. For the Lichfield translators, this was to deny a fundamental advance in botanical knowledge. In the controversy between Darwin and Withering about Linnaean plant sexuality, the Lichfield translators prevailed both in their terminology and in their underlying acknowledgment of the core of the Linnaean system. Theirs is the direction botany books took thereafter in presenting Linnaeus; the Lichfield translators were responsible for consolidating the terms "calyx," "panicle," "stamen," and "pistil" in botanical usage. By 1796 the sexualized and Latinized nomenclature was fully established. In the revised third edition of his *Botanical Arrangement*, Withering acceded to what had become established practice,

Frontispiece to Erasmus Darwin, *The Botanic Garden* (1789–91), drawn by Henry Fuseli.

included Latin terminology, and used Linnaean names for classes. "The English reader will perceive," he wrote, "that considerable changes have been made in the *Terms*, by a nearer approach to the Linnaean language; but in this point the Author rather willingly follows than presumptuously attempts to lead the public taste" (preface).

Between the publication of the Lichfield Linnaean translations during the 1780s and the third edition of Withering's Botany in 1796, Erasmus Darwin issued his notorious versification of the Linnaean sexual system, "The Loves of the Plants" (1789), which later formed the second part of his omnibus work *The Botanic Garden* (1791). "The Loves of the Plants" is a poem about nature, with lengthy informative footnotes. Through the voice of the "Botanic Muse," Darwin toured plant species, described the numbers of stamens and pistils in over eighty plants, and gave examples of all the Linnaean classes and orders. A prose account of the Linnaean sexual system prefaces the poem, and Darwin also discussed botanical matters in footnotes within the four cantos of the poem itself. The publishers of *The Botanic Garden* set a high price of twenty-one shillings for this work when they issued it in a two-volume quarto edition in 1791. Some years later Anna Seward, a poet and Darwin's younger friend, reflected on that moment: "The immense price which the bookseller gave for this work, was doubtless owing to considerations which inspired his trust in its popularity. Botany was, at that time, and still continues a very fashionable study. Not only philosophers, but fine ladies and gentlemen, sought to explore its arcana." [23]

"The Loves of the Plants" is not so much a systematic versified Linnaeus for the uninitiated as an evocation of the erotic impulse Darwin finds at the heart of the sexual system. The Linnaean sexual system serves as Darwin's point of departure for personified accounts of love and sexuality in the vegetable kingdom. His poem shows sexuality and reproduction as central preoccupations of the floral realm, motivating the behavior of the "males" and "females" in each flower. In rhapsodic verses, exuberant and sexualized, Darwin created brief narratives about plants and shaped verses about "Beaux and Beauties [who] woo and win their vegetable Loves." [24] He based his anecdotes on human analogy, transposed human sexuality into the plant kingdom, and wove poetic, highly visual accounts around a characteristic feature of a plant.

Darwin's declared purpose in *The Botanic Garden* as a whole was, he explained, "to inlist Imagination under the banner of Science," to bring scientific material to life and draw readers into scientific enthusiasms.[25] Darwin believed in Enlightenment ideals about science and improvement, supported increased school science for girls, and in later years included botany among the subjects in the ideal girls' school he described in *A Plan for the Conduct of Female Education in Boarding Schools* (1797). There he outlined a widely inclusive and progressive pedagogy, featuring science subjects and much more physical education than would have been common for those teachers and parents who pursued a feminine ideal of delicacy. Along with arithmetic, geography, natural history, mythology, and drawing, he recommends botany as a subject in itself and suggests specific books for introducing girls to botanical study. Darwin's *Plan* is at once a call for a wider curriculum for girls and a pedagogical guide to textbooks on the school subjects Darwin recommends as appropriate for young students and for girls.[26]

At the same time, Darwin entrenched sexual difference and gender differences in both his poetic and his pedagogical writing. "The Loves of the Plants" naturalizes a conventional sexual politics. His verses voyeuristically parade the virgin lily who "droops . . . with secret sighs" and the cowslip who "bows with wanton air" to "five suppliant beaux." We read about the Lychnis who "In gay undress displays her rival charms, / and calls her wondering lovers to her arms." Although some verses are sentimental, with tears, modest blushes, and maternal sighs, the overriding tone is the lascivious delight of a male-centered sexual fantasy. Darwin's ideas, like Linnaeus's, form part of eighteenth-century debates about sexual difference and gender differences. Although he made no portentous claims for his poem — he represents himself as "only a flower-painter" whose verses are "diverse little pictures suspended over the chimney of a Lady's dressing room" (proem) — Darwin's own images, for all his otherwise politically progressive views, are deeply polarized between the chaste, blushing virgin and the seductive, predatory woman. Although Darwin's poem gave a voice to female sexuality, his representations of women remain conventional, and conventionally polarized.

During the political turmoil of the 1790s, Linnaean plant sexuality became an emblem of questionable politics for those whose gender ideology excluded women from sexual knowledge or who equated sexual knowledge with practices discordant with their ideas about femininity. The year after the death of Mary Wollstonecraft, Richard Polwhele, an anti-Jacobin cleric and poet, directed the verse diatribe *The Unsex'd Females* (1798) at Mary Wollstonecraft and her "Amazonian band" of women writers. For political conservatives like Polwhele, committed to church, king, and country, Mary Wollstonecraft incarnated the dangers of freedom and revolution as associated with French republicanism and godlessness. The posthumous publication of William Godwin's biography of Mary Wollstonecraft, with its frank account of her personal struggles and unconventional life, made even some like-minded writers ill at ease about the tumult of her personal choices. She was an easy target for Polwhele, who censured her for leading literary women of the 1790s, such as Anna Barbauld and Charlotte Smith, away from the path of "feminine" sensibility. Polwhele had been taken up as a precocious poet by literary ladies in Bath, and his poem is fueled by indignation at how far some women had strayed from his feminine ideal, incarnated in the figure of the conservative and evangelical writer Hannah More. Polwhele framed his attack on "the female Quixotes of the new philosophy" in terms of an attack on female botanizing. He accused women who study plants of knowingly taking part in sexual activity and warned that if "botanizing girls . . . do not take heed . . . , they will soon exchange the blush of modesty for the bronze of impudence." One verse reads:

> With bliss botanic as their bosoms heave,
> [they] Still pluck forbidden fruit, with mother Eve,
> For puberty in sighing florets pant,
> Or point the prostitution of a plant;
> Dissect its organ of unhallow'd lust,
> And fondly gaze the titillating dust.[27]

Polwhele's linkage of pen and prurience radiates with cultural associations between the pen and the phallus, according to which the woman writer is being publicly, and transgressively, sexual.[28] Polwhele's footnotes in the poem make clear his objections not just to female botanizing but to coeducational botanizing by young people as well. Plant study becomes hands-on sexuality for Polwhele. His reactionary stance showed both prudery and prurience, as when he wrote that he "has seen girls and boys botanizing together." Polwhele's objections therefore combine critiques of female scientific practices with critiques of other "Gallic" and "revolutionary" practices, such as acknowledging sexuality and teaching children about sex.

Behind Polwhele's hysteria lies a debate about sexuality and national morality in the body politic, a topic of concern across the political spectrum during the 1790s. Mary Wollstonecraft herself had concerns about "impurity and chastity," and her fundamental argument in *A Vindication of the Rights of Woman* contrasts an ideal of human rationality, conjugal friendship, and virtue with the reality, as she saw it, of women improperly educated and hence easily enslaved to sensualists and to ideas about false refinement. Her conviction about encouraging female virtue and restraint did not turn her against teaching girls botany, however; she gave botany a place in the elementary curriculum she envisaged for her preferred coeducational day school.[29] The poet Anna Seward likewise had no difficulty with botany for girls; she dismissed those who complained that Linnaeus's language was unfit for a popular female readership. As she put it, Linnaean language of plant sexuality "can only be unfit for the perusal of such females as still believe the legend of their nursery, that children are dug out of a parsley-bed; who have never been at church, or looked into a Bible,—and are totally ignorant that, in the present state of the world, two sexes are necessary to the production of animals."[30] Pleasance Smith, wife of James Edward Smith of the Linnean Society, who then was writing his *Introduction to Physiological and Systematical Botany* (1807), remarked, apropos of Polwhelean attacks on botanizing women, "I shall be glad to see [botany], this beautiful and innocent study, rescued from all objection by *your pen.*—If nature may be admired and inquired into at all by women, surely *vegetable* nature is remote from indelicacy . . . , and it is much more likely to dispose the mind both innocently and religiously than the paltry pursuits to which so many are doomed. To the pure all things are pure, and to the depraved all objects animate or inanimate, natural or artificial, excite licentious ideas."[31]

In the "Botanical Conversation" from 1786, quoted at the start of this book, one of the women seems to take pleasure in mouthing botanical terminology, as though the use of technical terms is part of the lure of botanizing. There are complex cultural issues in the technical language of botany, but the issue of women and botanical language is, more importantly, part of a larger gender politics. The Linnaean system as a nomenclature in some ways facilitated botanical study, for Linnaeus introduced binomial names in place of the cumbersome diagnostic phrases then used for describing plants. The Swiss botanist Albrecht von Haller, Linnaeus's contemporary, had named one gentian *Gentiana corollis quinquefidis rotatis verticillatis*. By contrast to such an unwieldly descriptive sequence, Linnaeus added to the genus designation a one-word nickname for the species and renamed Haller's gentian *Gentiana lutea*. This species shorthand, first employed in Linnaeus's *Species Plantarum* (1753), greatly simplified identifying plants.

But while the spread of Linnaean nomenclature helped botanical discussion across national and linguistic boundaries, it also served as an exclusionary feature of botany. Latin botanical nomenclature introduced standardization, useful for commercial interests, for agricultural improvers, and for learned naturalists and scientists who wanted to distinguish their work from herbal vernacular and the local practices of an earlier generation. Linnaeus's system "epitomized the continental, transnational aspirations of European science, . . . and Linnaeus deliberately revived Latin for his nomenclature precisely because it was nobody's national language."[32] Nevertheless, those without knowledge of Latin or access to classical education were excluded from the inner sanctum of Linnaean botany. Latin was a gatekeeper, restricting access to those with the right credentials of sex and class. (The herbal tradition flourished deep into the seventeenth century despite regional and linguistic differences in nomenclature, and popular nicknames were the stock-in-trade of herb women, but popular views and learned views of the natural world diverged sharply during the late seventeenth century.) One woman in the Bluestocking circle, the elderly Frances Boscawen, showed her understanding of the politics of language in relation to botanical nomenclature when she wrote to her friend Mary Delany about a newly acquired plant for her garden:

> I bro[ught] from [Mr. Burrows] in my chaize a plant with a prodigious long Greek name, w[hich] I forgot before I got home, but the plant I hope . . . will take root and flourish. . . . its name begins with an M and is something like Mucephalus, but not just that; . . . it was new to me at least by its hard name, and if it has a soft one the nursery man wou'd not trust me with it, lest I shou'd despise the plant and its owner; both wou'd be more considerable in my eyes, he thought, for bearing and pronouncing so long a name.[33]

The difference between a "hard name" and a "soft name" for a plant carried status and significance for the nurseryman, and in all likelihood for her as well, and her comment suggests that both of them were aware of the struggle between "vulgar" and "refined" discourse that was a feature of the social and political history of language at that time.[34]

During the period 1760–1830, the simplicity of the Linnaean taxonomic system helped make botany a popular science in England, and the signs of that popularity were everywhere: in magazines, novels, popular books, verse, art, games, and public lectures. During these decades, English botany was largely Linnaean botany. Books proliferated, some intended for a general adult audience and others for audiences differentiated by age and sex. Botany began finding its way into school curricula, and many books were written for both formal and informal use, "in families and in schools." But by the 1820s different communities had emerged for studying plants. As part of their beliefs about nature, poets and protoscientists developed their own quite different ways of engaging with plants. Thus the Linnaean system of plant classification and nomenclature came under attack, for example, from romantic poets, essayists, and scientists who developed an approach to nature often guided by holistic and emotional concerns. Their antisystem spirit differed markedly from the taxonomic impulse of earlier botanists. For some romantics, botany was a science of words that led away from sensual apprehension of nature; Charles Lamb opposed informational approaches to nature altogether and wanted to replace Enlightenment botany and natural history with a new philosophical, aesthetic, and poetic program. Others saw the boundary between art and science as more porous. John Keats, for example, studied botany extensively as part of his medical training and drew deeply on his herbal and botanical knowledge for writing poems with a wonderful particularity; for him botany was one of the many faces of the vital impulse of nature.[35] Coleridge revered Linnaeus's contributions to botany but thought that a classificatory system based only on external characteristics could not lead to an understanding of truly causal laws. He aimed to define his relationship to nature through an interpenetration of art and science, a harmony of mind and nature.[36]

By contrast, botanical coteries of a more scientific kind took on discernible profiles by the 1820s. Some botanists turned to plant physiology as a new area of inquiry and sought to apply results to improving crop yields. Many botanists increasingly turned away from the Linnaean system of classifying plants according to their reproductive parts and grouped them into families according to a series of characteristics rather than by one diagnostic feature alone; they looked with interest to the "natural system" of plant classification developed by Bernard de Jussieu in Paris and Augustin-Pyramus de Candolle in Geneva. The change from Linnaean systematics to the natural system took place rapidly among specialists, more educated botanists, and those who read the higher-level journals. A review article in the *Annals of Philosophy* about advances in the sciences during 1819, for example, concluded by observing the progress made by the natural system and the increasing neglect of the Linnaean sexual system: "This release from the fetters of authority [the Linnaean system] cannot but augur good to the science; and we have no doubt, but that in a few years botany will be able to regain the time which has been lost in the arrangement of plants by the mere number, proportion, and connexion of their sexual organs, to the total neglect of the study of their affinities . . . [and] the systems of Rivinus, Tournefort, and Ray."[37] Teachers continued to explicate Linnaean botany for students, but increasingly it was seen as the gateway, or the lower rung of the ladder of botanical knowledge, associated with children, beginners, and women.

During the 1790s commentators began distinguishing between the "bota-nist" and the "botanophile," between the scientist and the enthusiast. These distinctions became entrenched over the next decades. They also became gendered, so that the botanist was male and masculine and the botanophile usually female and feminine. As a result, during the 1820s some botanists began to generate strategies to "defeminize" the public image of the science.

TWO

Women in the Polite Culture of Botany

The fair daughters of Albion have envinced a zeal and ardour in Botanical researches which have not only done the highest honor to themselves, but have eminently contributed to rescue these pursuits from unmerited reproach, to elevate them into reputation, and to impart to them, if not a superior value, at least a superior currency and fashion. — That such excellence should have been attained in this branch of science by so many of the female sex, notwithstanding the disadvantages they labour under from the want of scholastic and technical instruction, is a convincing proof of the liberality with which Nature has endowed the female mind.

Charles Abbot, *Flora Bedfordiensis*, 1798

The "Botanical Conversation" I began with, a representation from 1786, shows botany marked as appropriate for women and recommended within the dominant gender ideology of the 1780s. There we met Flora and Ingeana, two exemplary women who studied Linnaean botany and talked about the names of plants. Botany was a subject of enthusiasm for them, but when called they left the garden, put aside their botanical occupation, and returned to their other responsibilities as well-socialized women of their day. During this period botanical conversations took place among actual women in actual gardens. According to whether women were encouraged to study plants or admonished about the dangers of botany, their conversations were sanctioned or stigmatized. Increasingly, botany was a subject associated with women and hence was laden with larger, often unarticulated meanings.

From the 1780s on, botany accorded with conventional ideas about women's nature and "natural" roles. It fit the gendered assumptions of many writers and cultural arbiters about women and domestic ideology and was thought to be a way to shape women, or shape them better, for their lives as wives and mothers. Some writers, for example, saw botany as aesthetically concordant with feminine beauty or elegance or delicacy. Thus a correspondent in the *Gentleman's Magazine* claimed that the "nurture of exotics not only belongs more particularly to the female province, on account of its being an elegant *home* amusement, but because of there being much delicate work, essential to the welfare of plants, that is more dexterously performed by the pliant fingers of women, than by the clumsy paws of men. Ladies can also more conveniently attend the regulation of the green-house sashes, which require closer attention than the ordinary concerns of gentlemen will allow them to pay." [1] Others saw it, pedagogically, as a path to piety and health or as an antidote to superficial accomplishments. It was considered a "delightful pursuit," a "rational pleasure," a more suitable way for women "to court the fields, the woods, and gardens of the paternal seat, . . . more congenial to their nature, and more appropriate to their exercise, than barren watering places." [2] Others contrasted botany with studying animals or insects, pointing out that botany involved no killing, no cruelty, and no dissecting. Social commentators on botany in the late eighteenth century treated it as an activity for female leisure, for "amusement" and "recreation." Botany, in short, was a litmus test for other attitudes toward women.

During the later eighteenth century botany particularly became part of the social construction of femininity for girls across the middle and upper ranks of society. Like music, for example, botany was congruent with ideas about both gender and class and was linked to other polite activities in the lives of girls and women.[3] They studied plants as a fashionable form of leisure and as an intellectual pursuit rewarding in itself. This included collecting plants, creating herbaria, learning some botanical Latin, reading handbooks about Linnaean systematics, taking lessons in botanical illustration, using microscopes to study plant physiology, and writing introductory botany books. Queen Charlotte, wife of George III, embodied the popularity of botany as a conventional activity for women. Botany was her chosen pastime during the 1770s, 1780s, and 1790s. An encomium in the opening issue of *The Lady's Poetical Magazine* that represented Charlotte as an exemplary wife and mother also celebrated her scientific interests in these lines:

> Happy for England, were each female mind,
> To science more, and less to pomp inclin'd;
> If parents, by example, prudence taught,
> And from their QUEEN the flame of virtue caught!
> Skill'd in each art that serves to polish life,
> Behold in HER a scientifick wife![4]

Within the climate of the English Enlightenment, parents and teachers instructed pupils about science. They considered scientific endeavors far better than superficial social accomplishments for shaping the manners and morals of young women. Thus Queen Charlotte, joining an early interest in polite science with a love of flowers and gardens, received instruction in botany and ensured that her daughters would study botany as well. The king and queen were generous patrons of the botanic garden at Kew that George III's mother, Princess Augusta, and her adviser Lord Bute had worked so hard to create. In 1773 Sir Joseph Banks honored Charlotte by giving the name *Strelitzia reginae* to the exotic bird of paradise flower recently introduced into England, "as a just tribute of respect to the botanical zeal and knowledge of the present Queen of Great Britain." A herbarium that King George purchased for her was housed at Frogmore in Windsor Great Park, and the president of the Linnean Society was asked at one point to help her with it and to converse with the queen and the princesses about botany and zoology.[5] Amid the personal and political turmoil that buffeted Queen Charlotte during the 1790s, she made daily visits with her three eldest daughters to her botanical retreat at Frogmore; there, a contemporary reported, she "sits in a very small green room which she is very fond of, reads, writes, and botanizes."[6] When Charles Abbot, clergyman-botanist and fellow of the Linnean Society, dedicated his *Flora Bedfordiensis* (1798) to Queen Charlotte as "the first female Botanist in the wide circle of the British Dominions," he presented her as first among many other women at that time who also made their contributions.

Queen Charlotte's example undoubtedly gave a stamp of social approval to botany as associated with family and appropriate leisure and encouraged (and arguably was meant to encourage) women's entry into plant study. Women of the middle classes, gentry, and aristocracy joined the new workforce of polite botany as plant collectors, illustrators, and students of taxonomy. While their brothers and sons were away at school or on the Grand Tour for the luster of experience, gentlewomen on their estates and women of the new middle classes in town were at home, taking part in mandatory conventional social rounds. They were excluded from formal participation in the public institutions of botany and science. They could not be members of the Royal Society or the Linnean Society, could not attend meetings, read papers, or (with very rare exceptions) see their findings published in the journals of those societies. But writers, teachers, and sermonizers recommended informal participation in botany.

The Herbal Tradition

Long before the Enlightenment, knowledge of plants had been part of women's traditional work as healers. Peasants and aristocrats alike worked as "wise women" to apply their knowledge of plants in herbal medicine, midwifery, and medical cookery (the healing power of food). Their skills in plant lore, developed through experience and passed on in oral traditions, are examples of gynocentric science.[7] Herbal and healing traditions afforded women many opportunities to work with plants and develop knowledge and expertise. Herbals—illustrated volumes about plants that flourished with the spread of print—served botanical and medicinal purposes as handbooks for plant identification and as accounts of the culinary and medicinal uses of plants.[8] Medical remedies using herbs featured among the responsibilities of early modern gentlewomen such as Lady Grace Mildmay (1552–1620), who was well versed in herbal practices and operated "virtually as a professional doctor."[9] Women were active as healers, and as licensed practitioners, in various areas of medicine in the later seventeenth century. The poet and essayist Jane Barker, for example, studied "simpling" with her brother, gathered "many Patterns of different Plants, in order to make a large natural Herbal," and was so expert in herbal medicine that she wrote her "Bills in Latin, with the same manner of Cyphers and Directions as Doctors do; which Bills and Recipes the Apothecaries fil'd amongst those of the Doctors."[10] Male and female healers and empirics also included unlicensed, unorthodox quacks. More generally and more important, however, lay medical knowledge was widespread, and vernacular medicine was widely practiced. Within the climate of medical self-help, "kitchen physic" formed part of the domestic duties of women across a wide social spectrum, from the wisewoman of a village to the landowner's wife with medical recipe books in her kitchen.[11]

Familiarity with herbalism and women healers was taken for granted in George Farquhar's comedy *The Beaux Strategem* (1707), in which the wealthy and philanthropic widow Lady Bountiful dispenses medical advice and remedies to the poor. She "cures Rheumatisms, Ruptures, and broken Shins in Men, Green Sickness, Obstructions, and fits of the Mother in Women;—the Kings-Evil, Chin-Cough, and Chilblains in Children" and "has cured more People . . . within Ten Years than the Doctors have kill'd in Twenty." Lady Bountiful's seriousness about her work forms part of the humor of Farquhar's genial and humane comedy. Absorbed in assisting a character she believes to be ill, she fails to see the fakery masking flirtation and seduction. Yet she is not rendered foolish or made a butt of satire because of her herbal knowledge or her belief that she has "done Miracles about the Country here with my Receipts." Indeed, herbalism supplies the metaphor for the aims and methods of the comedy, the speaker in the prologue declaring, "Simpling our Author goes from Field to Field, / And culls such Fools, as may Diversion yield."[12]

By the 1740s, however, Lady Bountiful's herbal-medical knowledge belonged to a style of female medical practice soon to be superseded. Women's knowledge of "simples" could no longer be taken for granted. In *The Female Spectator* (1744–46), an important women's magazine filled with advice about female self-improvement, a correspondent recalled that "good house-wives" during his boyhood dried, preserved, and distilled herbs and used simples to perform "wonderful cures." "In these politer times," however, "such avocations . . . are beneath the attention of a fine lady." The writer declared: "Heaven forbid that, old as I am, I should render myself so obnoxious to the most charming part of the creation, as to advise them to return to that old-fashioned way of spending time." Although he "would not be by this understood to persuade ladies to turn physicians," he does suggest that women could benefit from cultivating some herbal knowledge.[13] To be sure, country women continued to use their knowledge of herbs, serving new economies. Sir Joseph Banks, a doyen of late eighteenth-century botany, recalled first learning his botany in the 1750s as a young pupil at Eton from the women who "culled simples" for sale to druggists and apothecaries.[14] But though lower-class women worked as community simplers in some areas, collected for the London markets, and sold to herb shops in Covent Garden, for example,[15] many women in the gentrifying middle ranks lost or left behind their familiarity with traditional healing practices and relied more on male physicians than on female skills. The success of John Hill's *Useful Family Herbal* (1754; 6th ed., 1812) illustrates a shift in style and authority by mid-century as male-linked "scientific" approaches supplanted earlier, more informal, and female-linked modes of healing.[16]

Botanical Art and Design

During the eighteenth century botanical illustration became a tool of both descriptive and systematic botany. Representations of plants were a centuries-old feature of books such as John Gerard's *Herball, or Generall Historie of Plants* (1597), with its many woodcuts to help distinguish them. Later illustrative works aimed for more botanical precision, particularly in depicting the parts of plants, and art became a form of botanical documentation. Early in the eighteenth century Elizabeth Blackwell melded the herbal tradition with artistic skill and produced *A Curious Herbal* (1737–39), a two-volume folio-size illustrated work "of the most Useful Plants, which are now used in the Practice of Physick." She drew, etched, engraved, and hand colored five hundred illustrations of plants and added descriptions and names in several languages, along with information about medicinal uses, drawn from a contemporary botanical text. She described parsnip, for example, as "more used in the Kitchen than the Shops, and . . . esteemed nourishing and a provocative to Venery." Yellow water lily "is accounted cooling and anodyne, and good in dilerious Fevers, and for the Heat and Sharpness of Urine, and all Kinds of Fluxes and Loosenesses."[17] Elizabeth Blackwell's herbal was compiled not for a specialized audience well versed in Latin intricacies but rather, as she explained in the introduction, for "such as are not furnished with other Herbals." Accessibility being a central feature of the book, all botanical information is in English.

Blackwell (ca. 1700–1758) produced her botany book to earn money, and her biographical circumstances, alluded to throughout the book and circulated in periodical publications of the day, gave her work a certain pathetic interest.[18] Her husband, the son of a Scottish academic family, was imprisoned as the result of a lawsuit, and she undertook *A Curious Herbal* to pay his debts. As daughter of a merchant family and niece of the professor of medicine at the University of Glasgow, she received support and encouragement from apothecaries, physicians, and botanists of her day for a project that might otherwise have been closed to her. She showed samples of plant drawings to noted physicians associated with the Worshipful Society of Apothecaries, based in the Chelsea Physic Garden in London, and they, along with Isaac Rand, a learned apothecary charged with care of the garden, promoted her work. Elizabeth Blackwell moved into housing near the Chelsea Physic Garden so as to receive fresh specimens for drawing. *A Curious Herbal*, issued as a serial publication, appeared in weekly parts. It enjoyed a high reputation in England and abroad across the eighteenth century; the *Herbarium Blackwellianum* (1750–60) was an enlarged and improved version with a Latin text, edited by Christoph Jakob Trew in Nuremberg.[19] Elizabeth Blackwell was one of very few women whose botanical work featured at all in the public record until the mid-eighteenth century. *A Curious Herbal*, among the earliest publications on botany by a woman, links an older female tradition of herbal work with an emergent female tradition of botanical illustration.

George Ehret was one of the best known among botanical illustrators of eighteenth-century England; his work is prized in our day for the austere beauty and precision of his flower portraits.[20] The son of a German market gardener, Ehret enjoyed the patronage of botanists such as Trew in Nuremberg, Jussieu in Paris, and Linnaeus himself and also was very busy and popular as a teacher of botanical art. During 1749–58, according to his own account, he taught "the highest nobility of England" how to paint plants and flowers.[21] His pupils were mostly female and mostly young: "two daughters of the Duchess of Portland, . . . two daughters of the Duke of Kent, the Countess of Carlisle, three daughters of the Earl of Essex, . . . Lady Carpenter, four daughters of Sir John Heathcote." From January through June each year, when aristocratic families were in town, he taught his pupils botany and botanical illustration. He emphasized careful observation and precision and called on his pupils to create art objects that were not merely ornamental.

Botanical art like Ehret's, existing at the junction of art and botany, suggests that the families of his female pupils sanctioned his mixture of the artistic and the scientific. Polite science melded with parental norms and with fashion. Drawing skills belonged to the roster of conventional accomplishments for girls of higher social class, and books of many kinds provided instruction. *The Lady's Drawing Book* (1753), for example, written by Augustin Heckle "to engage the FAIR SEX to a profitable Improvement of their Leisure Hours," taught how to draw flowers, showing how to move from a rough sketch to a finished drawing and then to a painted version. As examples it provided drawings of the flowers and stems of common plants; unlike the emergent tradition of genuine botanical drawings, it does not show the roots of plants, nor does it separate out the reproductive parts of the flower for special attention.

By contrast, Ehret's drawing classes for women express a cultural moment when interest in botanical exactitude overlapped with ornament and fashion. The 1740s and 1750s saw a floral mania in rococo dress design in England, and some depictions of flowers were remarkably naturalistic. Although ornate French patterns still were widespread, the silk designer Anna Maria Garthwaite (1690–1763) created something new in botanically realistic patterns that she sold to weavers in London. Her designs— over a thousand are extant—portray flowers in naturalistic sizes, shapes, and colors and even show the plants' roots. Garthwaite introduced into her designs ordinary plants such as the daisy and lily of the valley, but she also incorporated exotic flowers that were new introductions to England. Garthwaite, the unmarried daughter of a provincial clergyman, lived in London with a widowed sister and worked as a freelance fabric designer for several decades. During the 1740s she produced fifty to eighty designs a year, some commissioned by firms of mercers. She had family connections to an apothecary and a naturalist who belonged to an early botanical society, and her designs appear to be based on close observation of plants rather than on prints from flower books. Samples of her fabrics show that weavers followed her designs and produced botanical details on fashionable silk brocades and damasks for women's clothing.[22]

Anna Garthwaite's fabric designs illustrate the intersection of fashion with naturalistic style in art and design for both women and men. In 1741,

Floral fabric design from the 1740s by Anna Maria Garthwaite (fabric 59835). (Photograph courtesy of the Board of Trustees of the Victoria and Albert Museum)

for example, the duchess of Queensberry appeared at court in a gown that itself was a closely detailed landscape scene. Her friend and cousin Mary Delany described it as follows:

> The bottom of the petticoat [had] brown hills covered with all sorts of weeds, and every breadth had an old stump of a tree that ran up almost to the top of the petticoat, broken and ragged and worked with brown chenille, round which twined nasturtiums, ivy, honeysuckles, periwinkles, convolvuses, and all sorts of twining flowers, which spread and covered the petticoat; vines with leaves variegated as you have seen them by the sun, all rather smaller than nature, which makes them look very light; the robings and facings were little green banks with all sorts of weeds; and the sleeves and the rest of the gown loose, twining branches of the same sort as those on the

petticoat. Many of the leaves were finished with gold, and part of the stumps of the trees looked like the gilding of the sun.[23]

The fashion for hair ornamentation in London produced similarly elaborate coiffures. In addition to picturesque landscapes and a construction of the solar system in motion, hair styles incorporated real flowers, fruits, and vegetables; a headdress named the "kitchen garden" style called for attaching vegetables to side curls.[24]

During the decades after 1760, botanical drawing and flower painting continued at the nexus of art, botany, and female accomplishment. As women during the 1750s had taken lessons in botanical drawing from George Ehret, so women continued to cultivate flower painting and botanical illustration as drawing-room skills. They took lessons from botanical artists and worked from botanical drawing books. G. Brown's *A New Treatise on Flower Painting, or Every Lady Her Own Drawing Master* (3d ed., 1799), for example, taught how to hold a pencil and mix tints and included uncolored outline drawings of flowers to be filled in.

Mary Delany (1700–1788), a figure in aristocratic literary and political circles across eighteenth-century England, was an accomplished floral needlewoman and artist who embroidered flowers on clothing, bedcovers, hangings, and seat cushions and created flowers in shellwork and featherwork. In her early years she knew Swift and other Tory wits in London and Dublin and enjoyed many midcentury contacts with Lady Mary Wortley Montagu and Bluestocking circles. Later, as a genteel embodiment of earlier Queen Anne ways, she was taken into the family circle of George III and Queen Charlotte. She practiced the art of cut-paper work all her life, and in later years she turned her energies, and her declining eyesight, to creating "paper mosaiks." These were collages of flowers made from intricately cut and layered colored tissue paper. She created over nine hundred collages. Her "Flora Delanica" contains magnificent and strikingly lifelike representations of flowers, replete with subtle variations in color and with insect bites on leaves.[25] Mary Delany was much celebrated for this work. Erasmus Darwin, for example, commended her in "The Loves of the Plants" as part of his portrait of the papyrus. He wrote:

—So now DELANY forms her mimic bowers,
Her paper foliage, and her silken flowers;
Her virgin train the tender scissars ply,
Vein the green leaf, the purple petal dye:
Round wiry stems the flaxen tendril bends,

"Paper mosaik" by Mary Delany of *Volkameria lanceolata*, based on a drawing by Lady Anne Monson. (Courtesy of the British Museum)

Moss creeps below, and waxen fruit impends.
Cold Winter views amid his realm of snow
DELANY's vegetable statues blow.[26]

Sir Joseph Banks remarked that Mary Delany's collages were "the *only* imitations of nature he had ever seen from which he could *venture* to describe botanically any plant without the least fear of committing an error." [27]

In later years, with the spread of botanical activities, women in search of a livelihood could teach botanical art. Among a few women botanical artists who also taught drawing, Margaret Meen (fl. 1775–1820) came to London from East Anglia as a young woman and taught flower and insect painting, presumably to support herself as an artist. She exhibited at the Royal Academy during 1775–85 and later at the Water-Colour Society.

"Paper mosaik" by Mary Delany of *Passiflora laurifolia*. (Courtesy of the British Museum)

A letter of introduction presented in 1781 to William Curtis, botanist and author of the *Flora Londinensis*, praised her work as a painter.[28] The "most outstanding woman painter associated with Kew in the eighteenth century," Margaret Meen produced ten plates for *Exotic Plants from the Royal Gardens at Kew*, a projected serial publication that she planned to issue twice a year but that did not appear beyond its second issue.[29] Her contemporary Mary Lawrance (d. 1830) was so well known as a botanical artist that nurserymen such as James Lee sent specimens of their latest introductions for her to draw. "It was thought to be an honour for the owner as well as for the flower when Miss Lawrance painted its portrait." She exhibited her work at the Royal Academy from 1794 until her death. A folio monograph *A Collection of Roses from Nature* (1799), dedicated to Queen Charlotte, contains ninety hand-colored engravings. Another

London, May 1, 1799.

PROPOSALS

FOR PUBLISHING BY SUBSCRIPTION,

A COLLECTION OF

PASSION-FLOWERS

FROM NATURE.

BY MISS LAWRANCE,

TEACHER OF BOTANICAL DRAWING, &c. &c.

The Work to be etched and coloured to imitate Drawings, by Miss Lawrance, from the Originals now in her possession.

To contain every species of Passion-flowers, now in cultivation in the
English Gardens.

To be published in Numbers, each containing Three Species, and to be
comprized in Ten Numbers.

The Work to be printed upon a Superfine Wove Paper, 20 inches by 15,
and when complete will form an elegant Volume.

The Name and Botanical Description (according to the best Authorities)
~~Scatagraphy engraved on each Plate.~~ to be given in the last number

The Price to Subscribers will be Ten Shillings and Sixpence each Number
(to be paid on delivery); to Non-Subscribers the Price will be advanced.

Subscribers' names are received at Mr. Hookham's, No. 15, Old Bond-
street; Mr. White's, Fleet-street; Mr. Robinson's, and Mr. Clarke's, New
Bond-street; Messrs. Carpenter and Co. Old Bond-street; at each of which
places Specimens of the Work may be seen; and at Miss Lawrance's, No. 86,
Queen Ann-street East, Portland Place.

COLLECTION OF ROSES.

Miss Lawrance's Publication of Roses is now complete, and ready for
delivery, comprized in Thirty Numbers, and containing every approved Species now in cultivation in England, is printed upon a Superfine Wove Paper,
and is illustrated with an elegant frontispiece, and a descriptive Letter-press;
both of which will be delivered gratis, with the last Number. Price to Subscribers is Ten Shillings and Sixpence each Number, and may be had of Miss
Lawrance, No. 86, Queen Ann-street, East.

Mary Lawrance, "Proposals for Publishing by Subscription, a Collection of Passion-Flowers from Nature" (1799). (Courtesy of the Beinecke Rare Book and Manuscript Library, Yale University)

publication, *A Collection of Passion Flowers Coloured from Nature* (1802), issued by subscription, showed "every species of Passion-flowers, now in cultivation in the English Gardens." Mary Lawrance also gave lessons in botanical drawing, for which she charged "½ a guinea a lesson and a guinea entrance." A popular teacher for many years, she published a manual for students, *Sketches of Flowers from Nature* (1801).[30]

Patrons and Collectors

Women of the aristocracy participated in the social enthusiasms of collecting plants. Using their wealth to amass specimens, they also were botanical patrons. Increasingly women on great estates and even in more modest homes stocked their gardens, and later their glasshouses, with plants for botanical study and pleasure.

The paradigmatic aristocratic woman collector of the eighteenth century was Margaret Bentinck (1715–85), the duchess of Portland, only daughter and heiress of Edward Harley, the last duke of Oxford. Her marriage in 1734 to William Bentinck, second duke of Portland, brought her into a family with distinguished botanical ancestry, for the first earl of Portland had been superintendent of William III's gardens at Hampton Court. Margaret Bentinck had the resources to cultivate her extensive interest in botany and natural history, and as the duchess of Portland served these areas of study by her wealth. She opened her collections to the public, welcomed naturalists to inspect her holdings, and commissioned plant hunters to send her exotic specimens from all over the world. She received visits from many notable horticultural botanists and was a patron to botanists and botanical artists who came to her estate at Bulstrode Park in Buckinghamshire to catalog her plants and develop a pictorial record of her holdings. She was an important player in the networks of friendships and contacts that made up botanical and natural history culture at midcentury, and her largesse made a difference. The duchess of Portland sponsored John Lightfoot's journey to collect plants in the Highlands, and that clergyman-naturalist dedicated his *Flora Scotica* (1777) to her as "that great and intelligent admirer and patroness of natural history in general." [31] She also contributed to cultivating intellectual life more generally. She had a long-standing friendship with Elizabeth Montagu, who considered the duchess of Portland's estate at Bulstrode an ideal of social and intellectual life and took it as her model for gatherings of what became known as the Bluestocking circle in London. [32]

From the 1730s on the duchess of Portland botanized with her close friend Mary Granville Delany, creator of "paper mosaiks." Mrs. Delany's multivolume correspondence from the 1740s through the 1780s, a window onto people, events, and manners, records much of interest to social historians of botanical culture. In the 1760s the widowed Mrs. Delany, summering at Bulstrode, joined the duchess of Portland in botanical activities. John Lightfoot, who was often there cataloging the plant collection, taught them about Linnaean botany; she reported, for example, that he "goes out in search of curiosities in the fungus way, as this is now their season, and reads us a lecture on them an hour before tea, whilst her Grace examines all the celebrated authors to find out their [Linnaean] classes."[33] At the height of the season for mushrooms and fungi, Mary Delany and the duchess of Portland busily collected specimens. One episode illustrates their delight particularly well. Mrs. Delany set the following scene: "Her Grace's breakfast-room . . . is now the respository of sieves, pans, platters, and filled with [fungi], are spread on tables, windows, chairs, which with books of all kinds, (opened in their useful places), make an agreeable confusion; sometimes, notwithstanding twelve chairs and a couch, it is indeed a little difficult to find a seat!" Amid all this clutter of different kinds of mushrooms and fungi, the two women sit down to dinner, and Mrs. Delany's narrative continues: "First course ended—second almost—when said her Grace, looking most earnestly at the road in the park, with a countenance of dismay,—'A coach and six! My Lord Godolphin—it is his livery, and he always comes in a coach and six, take away the dinner . . . what will they think of all these *great puff balls?'*"[34] Mary Delany created many of her own intricate and beautiful flower collages at Bulstrode. She received specimens there from noted nurserymen and from the Botanic Gardens at Kew, for King George and Queen Charlotte instructed that she be sent specimens of any newly arrived foreign plants.

Across the century, plant collecting continued to be an enterprise for women of means. Lady Elizabeth Noel (1731–1801) of Leicestershire, for example, collected plants, compiled a Flora of Rutland, sent specimens for the Smith and Sowerby *English Botany*, and drew plants.[35] Among women of the landed gentry, Mrs. Egerton, wife of a wealthy and improving landowner in Cheshire, collected indigenous plants, had a fine herbarium, and established a botanic garden on her estate in the late 1770s. A correspondent of Sir Joseph Banks described her as "very fond of Botany, and very deep in it too"; her herbarium was "the neatest Hortus siccus that ever I beheld: all the Plants are laid down in the neatest manner and in the Highest Preservation."[36]

Botanical interests, particularly interest in systematic collecting and study of plants, spread across a somewhat wider class range after 1760. Lady Anne Monson (ca. 1714–76) is historically a bridging figure between aristocratic female plant collectors and the emergent fashion for botany among the middle ranks of society. A great-granddaughter of Charles II, she collected plants and insects, cultivated natural history study, and

Illustration of *Monsonia speciosa* (named in honor of Lady Anne Monson) in *Curtis's Botanical Magazine*, vol. 3 (1790), plate 73. (Courtesy of Hunt Institute for Botanical Documentation, Carnegie Mellon University, Pittsburgh)

developed proficiency in Latin. Within the polite culture of botany, Lady Anne contributed to the project of spreading botanical knowledge throughout the land. When she was in her mid-forties, during the late 1750s, she worked with James Lee on what became his *Introduction to Botany* (1760). She is said to have proposed to Lee that he undertake this pioneering popularization with its translated excerpts from the writing of Linnaeus. She is also said to have assisted him in this work, likely with the translations themselves, but she did not wish her work on this volume to be known. Lee, a busy nurseryman, acknowledged that he had much assistance in compiling his *Introduction to Botany*: "Were he permitted, [he] would readily acknowledge the Obligations he has to those who have kindly helped him in his Undertaking; but . . . some Injunctions oblige him to be silent on this Head." After a divorce in 1757, Lady Anne married a colonial army officer whose regiment was based in India and made several journeys there. She collected plants and insects in India and hired native artists to record Indian flora for her. In 1774, en route to Calcutta, she visited pupils of Linnaeus resident at the Cape of Good Hope and botanized with them; one botanist described her then as "a learned lady with a passion for natural history."[37] Linnaeus himself celebrated her by naming the genus *Monsonia* in her honor. Using a conventionalized hyperbolic language, he wrote: "I have long been trying to smother a passion which proved unquenchable and which now has burst into flame. This is not the first time that I have been fired with love for one of the fair sex, and your husband may well forgive me so long as I do no injury to his honour. Who can look at so fair a flower without falling in love with it, though in all innocence. . . . But should I be so happy as to find my love for you reciprocated, then I ask but one favour of you: that I may be permitted to join with you in the procreation of just one little daughter to bear witness of our love—a little Monsonia, through which your fame would live forever in the Kingdom of Flora."[38]

Linnaeus's Daughters

Between 1760 and the 1820s, women crossed the threshold into botanical culture in growing numbers, and botany became increasingly a feminized area, marked as especially suitable for women and girls. Like science more generally, botany was part of both education and recreation, particularly for girls, who, in line with the values of an emerging middle class, took up the handbooks and the hobby of collecting, observing, drawing, classifying, and naming plants. There was even a botanical martyr, Barbara Townsend Massie (1781–1816), an enthusiastic young plant collector in Cheshire who met with an accident during a botanical excursion and became an invalid for the rest of her short life.[39] During the Linnaean years, many women came to botany through their families and became helpmates and "fair associates" to fathers, husbands, or brothers. A pattern of family apprenticeship or mentoring is not unique to botanical or scientific culture but mirrors instead the connection through which women came into art or music.[40] Daughters of botanists often reaped the benefit of their fathers' interest in plants when serving as their companions or helpers or working within their networks.

When Linnaeus invited Lady Anne Monson into his fantasy of botanical paternity, his trope yoked rhetoric and reality. The paradigmatic daughter of a botanical father of the eighteenth century was Elisabeth Christina Linnea, the eldest daughter of Linnaeus himself. Linnaeus cultivated her interest in observing nature and opposed her learning "French or any other useless accomplishments," contrary to his wife's interest in giving their daughters a conventional education for their social class in Sweden.[41] In 1762, when she was nineteen years old, Elisabeth Linnea published a short report in the *Transactions of the Royal Swedish Academy of Sciences* on observations she had made in the garden of their home at Hammarby, near Uppsala. She had noticed, she wrote, an electrical or phosphorescent effect on nasturtiums at dawn and at dusk during the summer months and reported this to her father.[42] He had neither seen nor heard of it before and witnessed it himself. Linnaeus recommended that she submit an account of her observations to the Royal Academy. She accordingly described the phenomenon and offered a few conjectures about what might cause these flowers to appear to emit sparks. Might it be a reflection from the northern lights? Might it be in the eye of the observer? Couching her observations modestly, she then submitted them "to sharper eyes than mine." A nineteen-year-old woman would hardly assert bold claims in a public and written forum to which she surely had access only as the daughter of the famous scientist.[43] She presented herself as a handmaiden of science.

Elisabeth Linnea's story illustrates how paternal influence could enable and shape scientific work in botany while also setting limits on it and its directions. She married just two years after her report appeared, but the marriage was unhappy and she moved back to her parents' home with her young daughter. There are no further botanical traces, no evidence that she helped her father compile the multivolume herbarium that was the work of his later years.[44] When Linnaeus died in 1778, his will specified that his herbarium should be sold for the benefit of his daughters. His library, left to his son, was sold in turn on the son's death in 1783 because Linnaeus's wife and daughters were not "interested in keeping it."[45] Although Linnaeus did not object to Elisabeth's botanical interests as a young woman, he probably discouraged her as a young mother from further botanical work. Nevertheless, Elisabeth Linnea's work stayed in the public record. Linnaeus reported her findings in his *Species Plantarum*, and subsequent accounts of the history of eighteenth-century botany incorporated her observations. Erasmus Darwin celebrated her in his portrait of the nasturtium in "The Loves of the Plants." He wrote, "A faint-like glory trembles round her head," and "O'er her fair form the electric lustre plays," and his footnote describes the circumstances in which "Miss E. C. Linnaeus first observed the tropaeolum Majus to emit sparks."[46]

There were other "Linnaeus's daughters" in botanical culture at this time who entered the public record because of their fathers and became known for their contributions. In colonial North America, Jane Colden (1724–65), daughter of the first surveyor-general of New York, studied botany and described and identified local plants near the family home in the Hudson River valley. Her father was a fervent Linnaean who taught her the binomial system of nomenclature and translated Linnaean Latin terms for her. She in turn helped him compile a large list of local plants to send to Linnaeus. During the mid-1750s her father wrote about her to a distinguished European botanist: "If you think, Sir, that she can be any use to you, she will be extremely pleased at being employed by you, either in sending descriptions, or any seeds you shall desire, or dried specimens of any particular plant you should mention to me. She has time to apply herself to gratify your curiosity more than I ever had."[47] Cadwallader Colden's letter was meant to introduce his daughter, then an accomplished botanist in her early thirties, to European botanical networks. Nothing came of his efforts, or of the efforts by several other botanists who had benefited from her botanical collecting, to have Linnaeus name a plant in honor of Jane Colden. During her botanizing years she collected and described four hundred species of plants, made leaf prints and drawings of them, and produced a manuscript portfolio. She married in 1759, aged thirty-five, and died in childbirth seven years later. Now known as America's first woman botanist, Jane Colden had the support of her father and other male botanists and naturalists who acknowledged her botanical skills and also considered her a colleague. She produced one of the earliest local floras, but the manuscript was not published until 1963.[48] It appears that a well-intentioned father in the New World was unable to dislodge the likely reluctance of Linnaeus and his botanical colleagues in the 1750s to recognize and draw upon the scientific skills of a woman.

Within the eighteenth-century culture of botany, the artistic skills of daughters could be directed to family projects, as part of the "hidden enterprise" of women's work. Ann Lee (1753–90), for example, grew up at the heart of English botanical culture as daughter of the nurseryman-author James Lee and was inspired by her father "with a love of his own pursuits." She was given drawing lessons by Sydney Parkinson, the young artist hired as botanical illustrator for Captain James Cook's three-year

Silhouette of Anna Blackburne (1726–93), from *Profiles of Warrington Worthies*, 1854.

voyage to the Southern Hemisphere in H.M.S. *Endeavour*. Parkinson made watercolors of over 700 plants from among the 3,600 that the botanist Joseph Banks collected. He died on the voyage and left his painting equipment to Ann Lee in his will. During the 1770s Lee became an asset to her father and his friends by drawing plants in private collections. She was among the artists who were kept busy drawing exotics in the botanical garden of the Quaker Dr. John Fothergill in Essex, for example, with its 3,400 species of plants, many imported from North America. During the 1770s she also prepared a folio volume of drawings of the genus *Mesembryanthum*. Little else is known about her. She married, wrote a preface to her father's memoirs in 1810, and like Jane Colden, died in middle age.[49]

Unmarried daughters living at home could be fine companions for wealthy fathers with natural history interests. Anna Blackburne (1726–93), botanist, ornithologist, and patron of natural history, was the daughter of a widowed gentleman in the industrializing Midlands who had the means and leisure to cultivate his interests in natural history. Erasmus Darwin acknowledged her as "learned and ingenious" in his Linnaean *System of Vegetables*, and James Lee and others campaigned in the 1770s to have a plant named for her.[50] Anna Blackburne probably began learning about botany as a child during the 1730s and 1740s from general books on natural history in the library at the family seat, Orford Hall. In later years she and her father received visitors such as Johann Forster, a naturalist who taught during the 1760s at the nearby Warrington Academy, a boys' school well known among dissenters and freethinkers. She benefited from the connection to Warrington Academy, particularly from the school's openness to the exercise of natural curiosity. Forster frequently read his school lectures to Anna Blackburne.[51] She studied Linnaean botany and taught herself Latin so she could read Linnaeus's *Systema Naturae*.

Anna Blackburne's widowed father did not interfere with her intellectual interests, which gave him a companion and took her beyond conventional activities for women of her age and station. Had her mother been alive, the story of Blackburne's natural history efforts might have been quite different, for her mother might have guided her daughter toward pursuits more customary for their social standing. Novels in the eighteenth century are stocked with examples of the perils of motherlessness for young girls; it is mothers, they teach, who direct girls toward modesty as the proper path for proper womanhood. From the point of view of women's involvement with science culture, however, the record suggests that the motherless girl whose father had scientific interests and was willing to foster them in her was more likely to pursue science beyond polite accomplishments into some degree of learnedness and commitment.

In 1771 Anna Blackburne wrote to Linnaeus and offered to send him specimens of birds and insects she had received from her brother, "who lives near new York in north America, [and] who annually enriches my Cabinet with the productions of that Country." She then described her own history in acquiring scientific knowledge of plants: "I have labour'd under many difficultys, not knowing one word of latin when I began to study your Systema Naturae, which hath employed my leisure hours for the last 4 or 5 years, in which time tho I have not been able to acquire a thorough knowledge of it, I have so far suceeded as to be able to find out most things, wch: I assure you I think an ample recompence for all the pains it has cost me." Further displaying her zeal for learning Linnaean botany, she adds: "My Father has one of the best Collections of plants in this country, but at 73 years old, thought it late to begin to learn a new System and therefore I had little help from him."[52] Linnaeus already knew about Anna Blackburne, having heard, he writes, that "three botanical ladies . . . in 1769 disputed with and triumphed over the botanist in the (physic) garden at Oxford." Anna Blackburne's reply is brief and pointed: that the three ladies were one, and that one she herself, who dealt with a gardener, "a great dunce," who "was surprised to see a Lady that knew any thing of plants."[53]

A botanical wife also belongs in the eighteenth-century tableau of scientific companionship. Sarah Abbot assisted in the project that became the first published county flora in England. Charles Abbot's *Flora Bedfordiensis* (1798) lists over 1,300 "Plants as grow wild in the County of Bedford, arranged according to the System of Linnaeus." The clergyman-naturalist Abbot designated the book "for the amusement and instruction [of] the fair daughters of Albion." He remarked in the preface to this work on the female contribution to botany in his day and expansively praised the "constant superintending care of a fair Associate" toward the plants in his own garden. "To the same amiable and interesting partner of his pursuits and labours," the author continued, "he is also indebted for the preparation of a *Herbarium*. . . . But this is only one of the innumerable obligations, for which he is proud to acknowledge himself indebted to her assiduity and attachment." Sarah Abbot appears to have been responsible for a six-volume herbarium developed about 1790–1810; a historian of botany writing about Abbot's herbarium asks, "Should it more correctly be called Mrs. Abbot's?" She worked closely with her husband on his many botanical projects, which included sending in specimens for James Edward Smith's *English Botany* (1790–1814) and *Flora Britannica* (1800). Abbot is said to have been "persistent in his demands" that she be given credit for her contributions to these volumes.[54] (No credits were given to her by name.)

The absence of Sarah Abbot's name from her husband's *Flora Bedfordiensis* may have been a strategic choice on both their parts. As much as Abbot directed his *Flora* to "the Fair Daughters of Albion," he also positioned it in relation to the Linnean Society. In the preface he expressed the hope that the Linnean Society, "of which he has the honor to be a member, will excuse the mode in which [the work] now appears, in consideration of the motives, which induce him to consult as a leading object the convenience and accommodation of that Class of his Readers, who though comparatively and in the common acceptance of the term unlearned, may not and ought not on that account to be regarded as uninitiated, or as unenlightened." Charles Abbot was alert to the sexual politics of institutional botany in his day. Perhaps he worried that his attention to women readers would make his work appear less serious to institutional arbiters of the field.

Two decades earlier, when writing to Linnaeus, Anna Blackburne drew a telling distinction between two ways women of her day related to plants. There were those, she wrote, who were "very fond of plants" and others who "know them scientifically." By "scientific" Anna Blackburne meant being Linnaean, understanding Linnaeus's system for classifying and naming plants; by "scientific" she also meant knowing Latin. In the "Botanical Conversation" from the *New Lady's Magazine* of 1786 quoted in my prologue, Flora and Ingeana delight in earnest exchange about the plant they are identifying by its Linnaean name. Access to Latin nomenclature was a form of intellectuality in itself, a sign of a larger world to which some women wanted access (but from which women were formally excluded). Learning Latin was the equivalent of a puberty rite for boys during the Renaissance, and the later course of Latin teaching can be interpreted as continuing the psychological location of Latin as a system of education and socialization for boys.[55] Since serious access to Linnaean systematics was afforded only to those initiated into rituals based in male study of classical languages, how then did women cross the threshold into that knowledge system? When William Withering prepared the third edition of *Systematic Arrangement of British Plants* (1796), he added a "Dictionary of Botanical Terms," cross-referencing common English terms with the Latin botanical terminology Linnaeus developed for describing plants. His intended audience included women: "The ladies, too, who, in spite of the obstacles attendant upon a dead language, often have recourse to Linnaeus in the original Latin, will find their researches facilitated by [the dictionary]" (preface). While some women learned Latin and even studied Linnaeus in the original, others had recourse to handbooks and translations that helped them understand Linnaean nomenclature and classification. Letters to the *Gentleman's Magazine* during the 1780s and 1790s called for more botany books to be written in English rather than Latin. Concerning a proposed British pocket Flora, one correspondent wrote in 1797 to urge that it be written in English, to accommodate both sexes: "The British ladies are determined to excel those of every other nation as much in mental as they do in personal attractions. Among various pursuits, many of them have prosecuted the study of botany with an ardour and success scarcely to be credited, if we contemplate the difficulties which interpose to check the progress of those who are unacquainted with the dead languages."[56]

At the same time, a fear of female learnedness was a leitmotif in much eighteenth-century writing about women and science. Writers cautioned against women's learning or knowing too much—too much reading, too many languages, too much science. They phrased their cautions in terms of dichotomies and contrasted the woman who sought too much knowledge or the wrong sort with the woman who chose more appropriate arenas. The first was labeled pedantic, masculine, unmarriageable, and unmaternal; the second, the "properly feminine" woman, located her enthusiasms within her domestic concerns. The specter of female pedantry was waved in front of women and men, parents and daughters. During the 1780s when Thomas Martyn encouraged female botanizing through his botanical letters and promoted Linnaeus's "new language for botany that would spare the long periphrases of the old descriptions," he remarked, "Nothing is more pedantic or ridiculous, when a woman, or one of those men who resemble women, are asking you the name of an herb or a flower in a garden, than to be under the necessity of answering by a long file of Latin words that have the appearance of a magical incantation." [57] In that same gendered spirit, courtesy books warned against girls' pursuing "intricate theories and abstruse points of science." Ann Murry, for example, wrote in *Sequel to Mentoria* (1799) that technical terms and scientific expressions may be useful for the propagation of knowledge, but "their general adoption and familiar use would render a female, particularly one of tender years, subject to the reproachful epithet of pedantic, or rank her as a smatterer in learning." [58] Cultural prescriptions like those went hand in hand with shifting ideologies about mothers, mothering, and motherhood. Discourses of many kinds represented the "proper lady" as tidy, domestic, chaste, modest, delicate, and maternal, cultivating herself in order to serve others better. Writers applauded female knowledge that was harnessed to maternal and other family responsibilities and distinguished between appropriate kinds and degrees of female knowledge and excesses of female learnedness. Within those cultural codes, women botanized in the fields and at home, within families, and among friends. They used botanical interests and skills within print culture and featured botany in their poems, books, and essays.

THREE

Flora's Daughters as Writers during the Linnaean Years

To the Goddess of Botany

Of Folly weary, shrinking from the view
    of Violence and Fraud, allow'd to take
    All peace from humble life; I would forsake
Their haunts forever, and, sweet Nymph! with you
Find shelter; where my tired, and tear-swoln eyes
    Among your silent shades of soothing hue,
Your "bells and florets of unnumber'd dyes"
Might rest—And learn the bright varieties
    That from your lovely hands are fed with dew;
And every veined leaf, that trembling sighs
In mead or woodland; or in wilds remote,
    Or lurk with mosses in the humid caves,
Mantle the cliffs, on dimpling rivers float,
    Or stream from coral rocks beneath the Ocean's
    waves.

Charlotte Smith, *Elegiac Sonnets*, 1797

The expensive apparatus of the Observatory, and the labours of Chemistry, confine the science of Astronomy, and the study of Minerals to a few; whilst the research into the animal kingdom is attended with many obstacles which prevent its general adoption, and preclude minute investigation; but the study of Botany, that science by means of which we discriminate and distinguish one plant from another, is open to almost every curious mind; the Garden and the Field offer a constant source of unwearying amusement, easily obtained, and conducing to health, by affording a continual and engaging motive for air and exercise.

Lady Charlotte Murray, *The British Garden*, 1799

During the Linnaean years, women left their mark on botanical culture as collectors, patrons, artists, helpmates, and coworkers. Within the special writing configurations of eighteenth-century culture, they also were visible as botanical writers. Women tilled the literary soil of botany for pleasure, profit, and public acknowledgment. They wrote about botany in descriptive catalogs, children's literature, fiction, periodicals, didactic verse, and introductory books. There is a striking tradition of women writing textbooks in the form of letters and conversations. With notable exceptions, women's writing about botany contributed to the diffusion of knowledge rather than to its creation, and consisted of popularized works for children, women, and general audiences rather than books and essays for members of institutions like the Linnean Society or the Royal Society.

Women's writing had become a social phenomenon in England during the eighteenth century, as women embarked upon public authorship in unprecedented numbers. The tumult of the 1790s, as Polwhele's anger showed in his poetic diatribe *The Unsex'd Females*, included agitation about women authors and their writing. If Polwhele had his way the list of female authors, headed by Hannah More, would contain only those who spoke softly and hewed to his notion about gender-appropriate subjects. In 1798 he had a sizable list of women novelists, poets, and dramatists to vet for ideological acceptability. It is estimated, for example, that three hundred to four hundred women published books in the 1790s alone. Their repertory included religious works, polemical pieces, novels, children's literature, drama, and works on science.[1] But though women wrote in many genres, they stayed for the most part within limits on subject matter shaped by the gender ideologies of the day. Women novelists, as has been convincingly argued, found acceptance in writing communities of their day by telling their stories within formats considered to be women's domain. Hence they structured stories about love and marriage plots, featured families, and incorporated the didactic function of the moral tale into their own narratives. A pattern of enlargement of literary opportunity, then, was coupled with restriction of subject matter and form.[2] Many women writers, conforming to an ideology of the Proper Lady, muffled and camouflaged their voices.[3] Others were contestatory and subversive. Still others cultivated a rhetoric of accommodation. Indeed, the history of women's writing in the late eighteenth century can be read as a history of resourcefulness, as women writers worked within the conventions and codes of their day and developed strategies for attaining voice.

One notable strategy was the narrative voice of the maternal educator. Complex cultural discourses about mothers and maternity developed in England that endeavored to shape new mothers for new families, reforming women in order to reform the nation. Motherhood was elevated to sociocultural prominence. Discourses of the body constructed women as maternal instead of sexual, and high status was given to women as mother-educators.[4] One notable strand in the social construction of motherhood at that time was a female mentorial tradition in Georgian children's books, identified by Mitzi Myers, that features mother figures as rational educators.[5] This tradition is partly a narrative feature in juvenile writing and partly a progressive form of pedagogy. The maternal and pedagogical voice also is very audible in women's scientific writing of that time.

The Maternal Mimosa: Frances Rowden

From the 1790s through the 1820s, science writing often was set within a broad framework of self-improvement. It was particularly the case that books for women and by women articulated extrascientific reasons for study, as though knowledge for its own sake was not a defensible female pursuit, or at least not a publicly defensible one. Frances Rowden's *A Poetical Introduction to the Study of Botany* (1801) combined an emblematic tradition about women and plants with empirical knowledge and an interest in women's education. The book developed from the author's teaching at a girls' school whose curriculum included botany. Rowden set out to use Erasmus Darwin's "Loves of the Plants" as the basis for elementary botany lessons, but she found the language "frequently too luxuriant for the simplicity of female education." She therefore chose to recast Darwin's exposition of Linnaean plant sexuality into a form more suited to her target readership. The result is a bowdlerized account of the Linnaean system, a post-Polwhelean attempt to make botanical study respectable for girls. *A Poetical Introduction to the Study of Botany* combines scientific information with moral lessons. Rowden discusses the parts of a plant and schematizes Linnaean ideas, then illustrates the Linnaean system by a sequence of verses that depict the classes and orders. Personified narratives about each plant lead into moral teachings. Like other writers of elementary botany books at that time, Rowden had a larger agenda than botanical systematics for her Linnaean exposition. When deciding

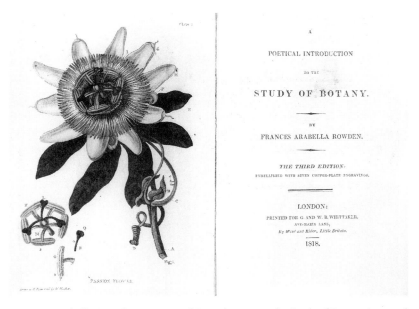

A

POETICAL INTRODUCTION

TO THE

STUDY OF BOTANY.

BY

FRANCES ARABELLA ROWDEN.

THE THIRD EDITION;
EMBELLISHED WITH SEVEN COPPER-PLATE ENGRAVINGS.

LONDON:
PRINTED FOR G. AND W. B. WHITTAKER,
AVE-MARIA LANE,
By West and Rider, Little Britain.

1818.

Frances Arabella Rowden, *A Poetical Introduction to the Study of Botany* (1801; 3d ed., 1818).

what to adapt from Darwin's "Loves of the Plants," she chose to include plant descriptions "from which some moral lesson might be derived, that the improvement of the heart might keep pace with the information of the mind."[6]

Rowden directs her *Poetical Introduction* to "young minds," to guide them "to attention and observation, and impress on their heart the beneficence of the Almighty." These are commonplace intentions in natural history books of her day, but Rowden has a particular interest in guiding minds that are both young and female. She believes that botany holds special benefits for women in the context of a prevailing domestic ideology, and explains herself this way: "As the situation of the female sex devotes [women] to a retired and domestic life, it is necessary they should acquire the great art of depending on themselves for amusement, and learn to concentrate their pleasures and pursuits within a narrow circle. — It is by such a regulation of their minds, that the foundation of a future happiness is laid, and they are enabled to contribute to that of others." Rowden has in mind activities that will yield self-discipline and moral training. Not all pursuits are equally valuable to this end, though, and she warns against those that are "immoderate and injudicious" or that are too sedentary. Botany, by contrast, improves the mind and also helps toward "a sound constitution" by drawing the female botanist away from the sitting room.[7]

Rowden teaches botany within a larger system of beliefs about female education, and her poems embody a deliberate gender ideology about women's place in the world. The personifications and the accompanying moralizing stanzas emphasize chastity, motherhood, and piety as well as male-female relationships that are familial, domestic, and nonsexual. To illustrate the Linnaean class Hexandria (six stamens), order Monogynia (one pistil), she versifies about lily of the valley (*Convallaria*) and evokes an emblematic tradition of female modesty and innocence. Thus the lily of the valley is meek and easily alarmed by attention; the flower, "the sweet emblem of the timid maid," "hides her blushing charms" in the shade. Other poems glorify mothers and maternal feelings. *Cypripedium*, lady's slipper, illustrates class Gynandria, order Diandria, in which two stamens grow directly out of the pistil. Rowden sidesteps this sexual image and describes the flower instead as a mother who "folds *two nurslings* to her downy breast." The poem ends with a historical exemplum of the "tender transports of maternal love," a story about the exiled Queen Margaret of Anjou, who withstood dreadful peril to protect her infant son. Rowden's gendered depictions teach both women and men about domestic, filial, and uxorial loyalty. Her personified flowers have innocent brothers, youths who are simple, noble, and dutiful, innocent shepherds, infants, or husbands who cannot wait to return home to "sweet domestic love." She features brothers in her poems who are "watchful" toward sisters who, meek and modest, are forever on the brink of being deceived and spoiled by the sexual perils of the world. Rowden illustrates the Linnaean category Syngenesia, Monogamia (plant has five stamens and one pistil) by the pansy, with "five fond brothers" who "round her trembling heart / Twine those strong bonds that Death can only part."[8]

Rowden's poems show a preoccupation with sexual threat toward modest virgins. This stands out in sharp relief when juxtaposed with Erasmus Darwin's verses on the same flowers in "The Loves of the Plants." The differences resonate, for example, in their dramatically different accounts of the *Mimosa pudica* (the "sensitive plant"), whose leaves close upon themselves and droop when touched. The mimosa is Darwin's example of the Linnaean class Polygamia, in which there are many stamens (the male part) and only one pistil (the female part). Darwin interpreted the species name *pudica* as "chaste" and orientalized the mimosa as a bride about to be introduced into a harem. On the brink of sexual knowledge, his mimosa is "alive through all her tender form," although "From each rude touch [she] withdraws her timid hands." He wrote:

Veil'd, with gay decency and modest pride,
Slow to the mosque she moves, an eastern bride;
There her soft vows unceasing love record,
Queen of the bright seraglio of the Lord.[9]

In contrast to Darwin, Rowden interpreted the species name *pudica* as "humble" and introduced an exemplary female figure that is neither orientalized nor overtly sexualized. Her verse reads as follows:

As some young maid, to modest feeling true,
Shrinks from the world and veils her charms from view,
At each slight touch her timid form receives,
The fair Mimosa folds her silken leaves;
. . . This tender plant to native feeling dear,
Sweet Sensibility delights to rear.

Frances Rowden's mimosa, a young and modest maid, is instead a model of the sentimental woman of feeling, guided by five guardian sylphs who, at her request, "soothe the passions of [her] breast." She reads as English and domestic, rather than a polygamous other. The moral derived from this account concerns the value of retiring from the dangerous and exotic world of folly into the world of maternal love:

So, my lov'd Pupil, may thy op'ning mind,
For Virtue's brightest deeds by Heav'n design'd,
In Nature's early, and Youth's tender hour,
Shrink from the blast of Folly's dang'rous pow'r;
And as thy step with modest grace retires
To scenes more suited to thy pure desires,
Maternal Love with rapture may behold
The ripen'd blossoms of her cares unfold.[10]

Rowden recasts Darwin's connubial verse with its suggestive sexuality into a paean to maternal bliss and the suppression of marital sexual feeling. She glorifies the maternal mimosa, in contrast to Darwin's male fantasy of female sexual awakening.

Is Rowden the Hannah More of women's botanical writing, another figure of backlash within early nineteenth-century culture? Her attachment to a prevailing domestic ideology and to a vision of female submission and maternality suggests that. Frances Rowden (ca. 1780 to ca. 1840) was a clergyman's daughter who worked as a governess in the family of Lord Bessborough and then became a teacher in London at a school for the daughters of fashionable families. She taught, and later was head-mistress, at the Hans Place School in Chelsea, 1801–20; she subsequently joined the French émigré coproprietor at a new school in Paris, and they later married.[11] Recollections by former pupils suggest that there was a gap between the public focus of Rowden's botany book and the private force of her teaching and personality, as though the public verve of her voice necessarily was moderated for her intended audience. The writer Mary Russell Mitford, for example, memorialized the sprightly and demanding Miss Rowden, who read her poetry, took her to the theater, and com-piled for her edification "abridgements of heraldry, botany, biography, mineralogy, mythology, and at least half a dozen 'ologies more." Rowden introduced her to Pope's Homer and Dryden's Virgil and "suffered [her] to admire Satan, and detest Ulysses, and rail at the pious Aeneas as long as I chose." At one point "Miss R." wanted her girls to stage Milton's *Comus* but received permission instead for "the only play fit to be acted by young ladies," Hannah More's pastoral drama *The Search after Happiness*. Mitford described the gusto with which Frances Rowden helped the girls shape something lively within a highly prescriptive play.[12]

The broad dilemma for women teachers and writers at that time was that, whereas Erasmus Darwin felt free to write verses about female sexual conduct in "The Loves of the Plants," his female counterpart would not have dared to do the same. The educator and poet Frances Rowden pub-lished *A Poetical Introduction to the Study of Botany* by subscription, under the patronage of clergymen and the families of her pupils, and accepted circumscribed themes and modes there (as also in subsequent publi-cations that combined moral teachings with literature).[13] It is notable, though, that despite personifications and moralizing stanzas, she taught botanical classification more directly than Erasmus Darwin did. Although the *Anti-Jacobin Review* criticized her for even taking Erasmus Darwin as a model, other reviewers praised the book as a "genteel guide to the science of botany." [14] While using science education as a vocabulary for teaching girls domestic and maternal ideologies, Frances Rowden also taught girls Linnaean botany.

Lady Charlotte Murray

Botany had social cachet as a fashionable science for women from the late Enlightenment into the mid-Victorian period. The Linnaean system for organizing the plant kingdom spawned many books that adapted his categories to British plants, among them *The British Garden* (1799). The book lists plants indigenous to Great Britain or introduced into British gardens from elsewhere, organized according to their Linnaean class and order. It also contains an introduction, addressed to "the young Botanist," that explains how "to discover the name of any unknown plant." The work combined, therefore, a descriptive catalog with an instructional account of Linnaean botany. Its author, Lady Charlotte Murray (1754–

THE

BRITISH GARDEN.

A

DESCRIPTIVE CATALOGUE

OF

HARDY PLANTS,

Indigenous, or cultivated in the Climate of

Great-Britain.

With their GENERIC and SPECIFIC CHARACTERS,

LATIN and ENGLISH NAMES,

NATIVE COUNTRY, and TIME OF FLOWERING.

WITH

INTRODUCTORY REMARKS.

VOL. I.

By the Rt. Hon. LADY CHARLOTTE MURRAY.

Second Edition.

PRINTED AND SOLD BY S. HAZARD:

Sold also by CADELL & DAVIES, Strand, LONDON; CREECH, EDINBURGH; and all other Booksellers.

M DCC XCIX.

Lady Charlotte Murray, *The British Garden* (1799).

1808), a noblewoman living in Bath, wrote to cultivate interest in botany and promoted botany as a more accessible study than other sciences.[15] Her exposition draws on Erasmus Darwin's translation of Linnaeus as well as "the able and elegant works of Messrs. Withering, Berkenhout, and other learned Writers on the subject." She does not trumpet her botanical expertise or areas of independent interest. Her intention was not "to enter into the endless and interesting discoveries which have been, and still continue daily to be made in this pleasing science." She frames her book in a way that is both conventional and specific to her social circumstances, describing it as written "at the request of a Friend." In the preface she distinguishes between guides on a journey, who actively describe the country and its beauties, and the direction post, which points out the road to the traveler and is "without any merit of its own." Her modest self-positioning was both historical commonplace and literary convention for women within the gender and class formations of the 1790s. Lady Charlotte Murray wrote within the polite culture of botany, and targeted an audience of young people, and others. Her book was well received as a "useful, convenient, and well compiled repertory," suited to "young botanists, and . . . dilettanti, while on a visit to Botanic gardens." One reviewer described it as not a book to "advance the science," but rather as one "calculated to foster and cultivate a love of this elegant pursuit."[16] *The British Garden* was directed to wealthy consumers of polite botany. Though it was issued in an expensive two-volume edition, the price did not hinder its sale, and editions continued to appear during the next ten years.

Botany for Profit and Solace: Charlotte Smith

Botany also was an economic resource for professional women writers. Charlotte Smith (1749–1806), widely published novelist of the late eighteenth century, capitalized on her botanical knowledge in natural history books for young readers and in verse. Her poems are steeped in technical knowledge of botany and filled with specificities. Embodying what Judith Pascoe has termed "a poetics of the botanically exact," they illustrate a new version of romanticism that now is challenging conventional accounts of canonical Wordsworthian romantic verse.[17] Smith also used the theme of women and botany to explore ways of being female in her world.

Charlotte Smith, having separated from her husband, embarked on a writing career during the late 1780s to support nine dependent children. She produced novels that mined conventional themes and gothic settings and demonstrated a sure satiric hand regarding lawyers, aristocrats, and the pretentious rich. She frequently criticized prevailing systems of female education in her novels and sketched smart, sensitive, and resourceful women. Her landscape descriptions are those of an attentive observer of leaves and trees, of furze and fern.

In the mid-1790s Smith added books for children to her repertory as a professional writer. Combining moral tales with natural history, she assessed the publishing possibilities and joined the company of Sarah Trimmer, Anna Barbauld, and Priscilla Wakefield in producing books for the juvenile natural history market. She began writing for this new market in 1794, at a time when one of her daughters was pregnant and dangerously ill.[18] *Rural Walks* (1795), one of several books meant to garner funds for her daughter's care, was written for girls twelve to thirteen years old. There she created the persona of a widowed mother who lives in forced retirement from the world because of a reversal in her fortunes. Mrs. Woodfield walks with her daughters and a city-bred niece in the countryside and converses about humility, charity, fortitude, the abuse of riches, and the absurdity of fashion. Like other books for girls from this period, *Rural Walks* is courtesy literature in a different generic guise and teaches girls how to be women. In an episode about gender and knowledge, Mrs. Woodfield, the maternal and mentorial narrator, introduces remarks about Mrs. Tansy, who "has been reading botanical books, till she fancies herself able to talk of such things to everybody, and worries one with something about petals, and styles, and filaments, and I know not what jargon." Mrs. Woodfield makes these comments as part of a conversation with her niece, a young woman raised fashionably and conventionally, who is meant to benefit from Mrs. Woodfield's lessons. Mrs. Woodfield explains that Mrs. Tansy shows off her botanical knowledge and turns to jargon because she does not understand botany yet wants to appear knowledgeable. As a result she is seen to be foolish and tiresome and is "driven back to her solitude," where "she can piddle about in her garden, and fancy she shall appear in print as a correspondent to a botanical society."[19]

In line with Smith's progressive and feminist orientation, the remarks about Mrs. Tansy in *Rural Walks* do not veer into condemning female aspiration to knowledge. The exemplary Mrs. Woodfield criticizes her for showing no judgment in conversation and for choosing an inappropriate social forum for displaying her knowledge. But Mrs. Woodfield sympathizes with the direction that Mrs. Tansy's "love of fame" has taken, for she too has botanical enthusiasms. Mrs. Woodfield has learned, however, to check herself frequently, "remembering how many other things I have to do, more material than considering of what genus a flower is, and what are its characters." The end of the dialogue makes clear that Mrs. Woodfield, a woman in the female mentorial tradition, prefers scientific conversation to gossip; it is "much better to talk of rhododendrons and toxicodendrons, merispernum and cenothusas, and other hard-named plants" than to be preoccupied with small talk.

Mrs. Woodfield is Charlotte Smith's representation of a conventional and acceptable way for women to harness botanical enthusiasms to maternal responsibilities and gender ideology. Smith did just that herself. She invoked the "Goddess of Botany" in her sonnet of that name in the collection *Elegiac Sonnets* (see the epigraph to this chapter) and drew on her own botanical knowledge in her writing. Sections in her unfinished poem "Beachy Head," for example, read like the verse notes of a naturalist and reflect a poet, "an early worshipper at Nature's shrine," who carefully observed hedgerows "Where purple tassels of the tangling vetch / With bittersweet, and bryony inweave."[20] There, as in other poems, Smith included botanical footnotes and integrated botanical terminology and Linnaean plant names. Explicit botanical references also thread through her novels. In *The Young Philosopher*, for example, the freethinking hero, who educates his sister to be "a rational companion," teaches her "to draw scientifically." His heroic female counterpart is a mother linked to America and freedom, who is "passionately fond of plants" and who teaches her daughter astronomy and botany.

In 1798, the year *The Young Philosopher* appeared, Charlotte Smith acknowledged a suggestion from the president of the Linnean Society, James Edward Smith (no relation), that she turn her botanical knowledge to literary and financial advantage. Her letter to him reads as follows: "I have not forgotten (being still compelled to write, that my family may live) your hint of introducing botany into a novel. The present rage for gigantic and impossible horrors, which I cannot but consider as a symptom of morbid and vitiated taste, makes me almost doubt whether the simple pleasures afforded by natural objects will not appear vapid to the admirers of spectre novels and cavern adventures. However, I have ventured a little of it, and have at least a hope that it will not displease those whose approbation I most covet." (She also reported that "after a long and successless struggle" she was compelled to leave England for the Continent, where at best she might be able to "botanize on annuals in garden pots out at a window." Nevertheless, she declared, "my passion for plants rather increases as the power of gratification diminishes.") [21]

A year earlier, when writing *The Young Philosopher*, Charlotte Smith had enlisted help from James Edward Smith for a botanical project that she hoped would yield financial reward. She proposed to her publisher a book to be done jointly with her sister that would combine illustrations of the Linnaean system with botanical information "for the use of botanical students & those who cultivate this branch of drawing." She explained in an unpublished letter from August 1797, "I am sure the thing wd answer because the plan does not interfere either with Sowerbys or Curtis's & would be the desirdrata [*sic*] for those who wish to study the elements of botany without distracting themselves with the terms that are so alarming to beginners or going to the expense of buying elementary books, where there are seldom any thing more than very bad plates copied from the same models time out of mind." She added that "Dr. James Edward Smith would correct it for me." This project did not reach fruition.[22]

But Smith continued to seek ways to fold her botanical interest and expertise into her writing, most notably into *Conversations Introducing Poetry: Chiefly on Subjects of Natural History* (1804). Written "for the use of Children and Young Persons," the book consists of poems about insects, quadrupeds, birds, marine life, and plants that sit inside conversations and commentary on many subjects. Many poems in this collection reflect on plants. "Wild Flowers," for one, cites many kinds of flowers, and Smith attaches two pages of notes to the poem with Latin botanical names of plants. In a more emblematic vein, "The Hot-House Rose" compares hothouse and garden specimens and reflects on the effects of nature and art. Similarly, "The Mimosa" names various types of mimosas. Smith's mimosa is an "emblem of excessive sensibility," and the mentorial figure indicts the plant for "fastidious pride" and "affectation." As in Smith's first book for children, Mother is the teacher whose domain includes moral values and cultural attitudes. Mrs. Talbot seeks to give her children "a taste for those pure and innocent enjoyments that are always to be found, which not only amuse the passing hour, but teach us, 'To look thro' Nature up to Nature's God.'" This means preferring natural history to playing billiards or reading bad satires, romances, or supernatural horrors, and it particularly means giving attention to plants.[23]

"Flora," the capping poem to *Conversations Introducing Poetry*, is Charlotte Smith's version for girls of Erasmus Darwin's *Botanic Garden*. Carrying many botanical footnotes, the poem describes Flora descending to earth in spring and touring among the flowers, trees, shade plants, and marine plants that make up her kingdom. The poem is meant to inspire interest among the girls in Darwin's "children of imagination" by presenting a "little cabinet picture of Flora." Verses introduce the goddess Flora and the sylphs and nymphs who attend her and "guard the soft buds, and nurse the infant flowers / . . . And save the Pollen from dispersing wind." Smith's poem, not principally a didactic and expository poem about botany, plays with botanical knowledge, as when Smith names Flora's nymphs for parts of the flower, "Petalla," "Nectarynia," and "Calyxa." As a version of *The Botanic Garden* for girls whose reading has not prepared them for the sophistication of Darwin's verses, "Flora" sidesteps plant sexuality and marriage in the plant kingdom; the goddess Flora is "simply clad" rather than surrounded by the splendors of Darwin's "highly coloured description."[24]

Charlotte Smith's literary career illustrates the dense interweaving of women, gender, and botanical culture in England at the turn of the nineteenth century. In the concluding conversation of *Conversations Introducing Poetry*, Mrs. Talbot's daughter Emily reports an unhappy exchange she evoked when, in the company of two fashionable young women, she named the flowers in a nosegay:

> "Lord," cried one of them, "what signifies what the name of this plant and that plant is? . . . I am sure I should never know one flower from another if I was to live an hundred years, and I wonder what good it does?"—"O! 'tis the fashion, you know," cried a Miss . . . in a drawling tone—"but if *my* existence depended on it, I'm sure *I* could never make any thing of remembering these hard names."[25]

The mother-educator responds by encouraging female botanizing, while insisting that women should observe certain proprieties about when to express their botanical learning in public and with whom. In both *Rural Walks* and *Conversations*, Smith presents the solitary pleasure of botany for a woman as less valuable than using botany in a family or social context.

Although the mothers in her children's books suggest the conventional pose of a professional writer, Smith's many pronouncements about botany make evident her own engagement and delight in botanical pursuits. Charlotte Smith, genteely raised, learned the skills of her class. She showed talent for drawing as a young girl, had the benefit of private lessons with a landscape painter, and produced watercolors of plants and flowers throughout her life.[26] Necessity turned her observational skills and botanical enthusiasms into her writing. In *The Young Philosopher*, the botanical mother recalls a time of great sadness and isolation in her life when she lived in a rustic house whose only adornments were mosses and lichens: "Amidst the many sad hours I have passed, I have never failed to feel my spirits soothed by the contemplation of vegetable nature."[27] Botany held considerable value for her as a woman writer with a complex personal story, who successfully integrated botanical expertise into popular novels, juvenile natural history books, and poetry.

Henrietta Moriarty

For women writers who wanted to avoid affronting public taste, the unfamiliar language of Linnaean botany and the visual portrayals of plant sexuality created dilemmas. Drawings of plants by the famous Linnaean botanical artist George Ehret highlighted the reproductive parts of a flower, and pictorial representations conventionally showed the stamens and pistils of plants in separate drawings in a corner of each botanical plate. This was a problem for Henrietta Maria Moriarty (fl. 1803–13), whose *Viridarium* (1806) consists of illustrations of flowers arranged alphabetically by their Latin botanical names. She explained the difficulty in her preface:

> Those who have the instruction, or, I might say, the formation, and even the fashioning of young minds most at heart, often find it difficult to obtain representations in this most pleasing branch of natural history, on the one hand sufficiently accurate, and on the other, entirely free from those ingenious speculations and allusions, which, however suited to the physiologist, are dangerous to the young and ignorant: for this reason I have taken as little notice as possible of the system of the immortal Linneus, and of all the illustrations and comments on it; nay, I have not once named the fanciful Doctor Darwin.

To avoid the "ingenious speculations and allusions . . . dangerous to the young and ignorant" of plant sexuality, Henrietta Moriarty chose to omit separate drawings of the reproductive parts of flowers from her book. *Viridarium* combines illustrations and Linnaean descriptions of plants, and also supplies information about the cultural requirements of plants.

A how-to book for readers who have no gardeners to turn to for advice, it includes popular greenhouse specimens of the period, such as the caper plant. Henrietta Moriarty labeled her book suitable for "the rising generation" and particularly for young ladies, and the second edition, issued under the new title *Fifty Green-House Plants* (1807), carried the subtitle "for the Improvement of Young Ladies in the Art of Drawing." The new edition promoted the illustrations as model drawings "[for use] in public boarding-schools" and partly as a way to teach botany.

Henrietta Moriarty made a tactical decision to exclude from her book botanical information that some readers (and some buyers) might construe as inappropriate. A colonel's widow with children, she wrote out of financial necessity and likely was a teacher of "the rising generation" herself, perhaps as a governess in a noble family.[28] Her writing career embraced two editions of her illustrated botany book, as well as novels whose narrative themes point to the marital and economic realities that led many women to look to writing as a plausible route to self-sufficiency. *Brighton in an Uproar* (1811) centers on Mrs. Hubertine Mortimer, needy gentlewoman and mother of four children, recently widowed after an unhappy marriage. The plot turns on her precarious situation as an unprotected woman in a predatory society, and the novel is filled with a sense of grievance about the unjust treatment of women in business matters. At one point in the novel Mrs. Mortimer enters the pencilmaking trade. At another, she is advised to write an illustrated botany book and publish it by subscription. Her adviser even recommends that she issue her illustrated botany book in two versions, one being "a small epitome, that the subscribers might have their choice."[29] Henrietta Moriarty herself marshaled her botanical accomplishments and converted the gendered skills of drawing and novel writing into an economic resource for herself.

Discourses of Utility: Sarah Hoare

By the 1820s, when Sarah Hoare issued books of botanical verse, botanical writing was an acknowledged resource for women who endeavored to secure a livelihood by their pens. Sarah Hoare (1767–1855) was an aging teacher when she published *A Poem on the Pleasures and Advantages of Botanical Pursuits* (1826) and *Poems on Conchology and Botany* (1831). A member of the Society of Friends, she exemplified the Quaker call for members to study nature as a salutary amusement and as a source of spiritual knowledge. Hoare belonged to the notable Quaker network of Hoares, Gurneys, and Barclays who combined banking and business with philanthropy and antislavery work at the turn of the nineteenth century. She did not marry, and she lived in Ireland for many years teaching the daughters of Quakers. By 1815 she was back in Bristol, where she continued to teach the daughters of Friends.[30]

Sarah Hoare first entered the public literary culture of botany with "The Pleasures of Botanical Pursuits," a verse encomium that was tipped into the eighth edition of Priscilla Wakefield's popular and influential textbook *An Introduction to Botany* (1818). There the poet, identified as "S. H.," celebrated her youthful pleasures in botany: "O Botany! the ardent glow / Of pure delight to thee I owe, / Since childhood's playful day." The poem touches on the aesthetic, medical, and spiritual benefits that come from pursuing "nature's rich array" and addresses Linnaeus: "Linné, by thy experience taught, / And ample page so richly fraught with scientific lore, / We scan thy sexual system clear." Verses describe various flowers, such as the calla lily and the pimpernel, using Linnaean terminology and nomenclature. Sarah Hoare's poem is not Linnaean systematics in verse, such as we find in Erasmus Darwin or Frances Rowden. Although she personifies flowers in the manner of "The Loves of the Plants" and *A Poetical Introduction to the Study of Botany*, her personifications are not tools for teaching systematics, morals, or lessons in domestic ideology. Instead, Hoare emphasized the medicinal uses of plants, such as the "soothing lymph" of the poppy. The utility of plant knowledge became an increasingly prominent theme in botanical writing during the early decades of the nineteenth century as writers studied the practical application of botanical knowledge to agriculture, medicine, cooking, and other household areas.

The uses of botany were her focus in *A Poem on the Pleasure and Advantages of Botanical Pursuits, with Notes, and Other Poems* (1826). Verses emphasize a medicinal approach to plants rather than tracing the moral, taxonomic, or emblematic route of other writers. Hoare wrote:

> For not alone to please the eye,
> Or deck our fields, this rich supply of ornaments profuse;
> Medicinal their juices flow, Nor idly do their colors glow,
> For He, who dress'd the beauteous show
> Assign'd to each its use.[31]

Written by "a Friend to Youth, Addressed to her Pupils," Sarah Hoare's poems about the utility of botanical knowledge were also linked to personal utility. At a late stage in her life she found herself "obliged to make every exertion in my power for a subsistence, and to secure the means of preserving myself from indigence, or becoming a burden to my friends, should I live to experience the helplessness of old age" (preface). The same spur to publication probably accounted for a subsequent publication, *Poems on Conchology and Botany* (1831), that reprinted her botany poem, along with a new poem about shells, grouped according to Linnaeus's arrangement.

Sarah Hoare's attention to the utility of plants has a notably gender-specific coloring. In addition to making a general point about the moral and religious benefits of female plant study, she also articulates a more specific theme about mothers and domestic duties. The Preface to *Pleasures and Advantages of Botanical Pursuits* abounds in discourses of maternity. Hoare addresses former pupils who had become mothers, warns them against neglecting their domestic duties and assigns them central responsibility for forming the character of their children.

She venerates the opinions of the conservative and evangelical Hannah More, author of the much-reprinted *Strictures on the Modern System of Female Education* (1799), who, opposing Wollstonecraftian radical arguments about women's education and place in society, insisted on differences in sex, gender, and class as foundational features of a secure society. She cites More's definition of female education as that "which inculcates principles, polishes taste, regulates temper, cultivates reason, subdues the passion, directs the feelings, habituates to reflection, trains to self-denial, and more especially that which refers all actions and feelings, sentiments, tastes, and passions to the love and fear of God."[32] Like Hannah More, Sarah Hoare endeavored to shape mothers. Her language on this subject is double-edged. Referring at several points to "the happy effects of the management of worthy mothers," she also declares that "the good of society in general principally depends upon the judicious management of mothers." Hoare sought through her writing to inculcate an ideology of motherhood that will "manage" the mothers as well as train them for managerial tasks in relation to their own children. In this regard, she places botanical knowledge firmly into the educational repertory of mothers and prescribes teaching botany as part of maternal responsibility.

The cultural campaign to redirect women toward motherhood and family spanned the political spectrum of England during the late eighteenth and early nineteenth centuries. Sarah Hoare's constructions of maternity point to the complex discourse in this area. Many women writers made maternal and domestic ideology work for them as social, intellectual, and economic resources. Mothers and teachers had a voice and claimed authority in schoolrooms and on the printed page. Women writers shaped the maternal and teacherly voice into a powerful tool in the mentorial tradition, and botanical culture was a beneficiary of an enlarged role for women in scientific education.

FOUR

Botanical Dialogues: The Cultural Politics of the Familiar Format

The spirit of improvement, that distinguishes this enlightened age, shines in nothing more conspicuously than in education. Persons of genius have not thought it unworthy of their talents to compose books purposely for the instructions of the infant mind and various ingenious methods of facilitating the acquisition of knowledge have been invented.

Priscilla Wakefield, *Mental Improvement, or The Beauties and Wonders of Nature and Art*, 1794–97

Here, take the plant in your hand, while I read what Withering says about the genera.

Sarah Fitton, *Conversations on Botany*, 1817

From 1780 to 1830, two generations of women writers in England gave a female imprint to popular science books in what was called the "familiar format." Using letters and conversations as narrative forms, they featured families, home-based informal settings, and maternal teachers. Women writing about science in the familiar format integrated topics of general interest and shaped a geography of domestic settings and family routines. Their pioneering contributions to the textbook tradition merged gender and genre and helped make science into popular and improving hobbies for the middle ranks of society.

Popular science books are part of the generic range of women's writing and belong to the still largely undocumented history of women's science writing.[1] Among early books written by women about chemistry, natural history, conchology, entomology, and mineralogy, popular introductory botany books constituted the largest group.[2] During the seventeenth century, women's writing about plants had consisted of books about herbal cures for the medical remedies market. Thus Elizabeth Talbot Grey, countess of Kent, wrote *Choice Manuall, or Rare and Select Secrets in Physick and Chyrurgery* (2d ed., 1653), a medical receipt book and self-help book for women, with special emphasis on women's illnesses; and Hannah Woolley's *Gentlewoman's Companion* (1675) contained home remedies, including "Salves, Oyntments, Waters, [and] Cordials" that she had learned to make herself as a wife and governess.[3] By the late eighteenth century, gender ideologies about women as writers and readers contributed to shaping a different kind of botany book by women and for women, and most botanical writing by women in this period took the form of introductory books written in the familiar format.

Letters and conversations had special prominence as literary technologies in eighteenth-century England. The epistolary form and the dialogue or conversation, touchstones in literary and cultural history, were foundational narrative modes for fiction, pedagogical writing, and popular science writing, and also underlay the nascent advertising industry of commercial culture.[4] Within science culture, letters were widely used for science writing and scientific communication in the early modern period—in writings by Galileo and Boyle, for example.[5] They were an excellent form of scientific instruction for those living in relative isolation from centers of science culture and those with little or no access to public institutions of science and scientific learning. Letters brought science to aristocratic pupils or to mothers and children on their country estates, surrounded by nature yet lacking the training to study it. Letters also were an acknowledged way to bring neophytes into science. Thus William Spence and William Kirby, authors of *Introduction to Entomology* (1800), looked to the letter form as a way to make their favorite science more pleasing and attractive; one of them wrote that the form "admits of much latitude in amusing digressions. . . . I know how much more pleasant it is to have the at best dry materials of terminology conveyed in a familiar style, and made palatable by an attractive vehicle." The letter became their chosen narrative form, the rhetorical means of drawing readers over the threshold and into the "vestibule" of entomology.[6] Within botanical culture, the epistolary format became one important convention for teaching women and young pupils about botany; Rousseau used letters to teach botany to his friend Madame Delessert, and Thomas Martyn continued the format in his *Letters on the Elements of Botany* (1785).

The dialogue or conversation also had a rich heritage as a narrative of accessibility. Bernard de Fontenelle's *Entretiens sur la pluralité des mondes* (1686) and Francesco Algarotti's *Il Newtonianismo per le dame* (1737), both available in English translations in their day, created conversations between a well-versed philosopher and a suitably curious marchioness as the way to teach polite audiences about the telescope, the microscope, Cartesian vortices, and Newtonian science.[7] Conversations like theirs belonged to the culture of sociability and were among the protocols of eighteenth-century polite culture. Reformers elevated conversation to the list of approved activities, contrasting it with card playing, for example. Thus when Hester Chapone, Bluestocking and educator, promoted serious conversation as a route to virtue, she proffered the opinion that conversation influences "our habits of thinking and acting, and the whole form and colour of our minds, . . . that which may chiefly determine our character and condition to all eternity."[8] To that end, manuals taught rules and practices of polite conversation and literary works illustrated good and bad paradigms of the art.[9]

By the late eighteenth century, women writers of educational books appropriated conversational narratives for their own purposes. Mary Wollstonecraft, as a progressive educationist who believed that teaching and learning should be grounded in everyday individual experience, used the dialogue form in her fictional work *Original Stories from Real Life* (1788). Priscilla Wakefield's *Mental Improvement, or The Beauties and Wonders of Nature and Art* (1794–97) constructed conversations between parents and their children about topics in natural history and interleaved discussion about bees and salmon roe with moral content. Maria Edgeworth, sharing Wollstonecraft's progressive pedagogical principles, explained in her book *Practical Education* that she had learned from teaching young children that "an early knowledge of the first principles of science may be given in conversation."[10] Many writers shared that view and adopted the conversational form for expository science. Jane Marcet's first book of popular science, *Conversations on Chemistry* (1806), consists of conversations between "Mrs. B.," a teacher, and two female pupils. Marcet explained in the preface that she considered the familiar format helpful for "the female sex, whose education is seldom calculated to prepare their minds for abstract ideas, or scientific language." She continued to use the conversational form for science writing in other titles, such as *Conversations on Political Economy* (1816), *Conversations on Natural Philosophy* (1819), and *Conversations on Vegetable Physiology* (1829), and her pedagogical narrative team of Mrs. B., Caroline, and Emily became the model for subsequent authors.

The "familiar format" was the conventional, indeed the canonical, form for most botany writing by women from the 1790s through the 1820s. The narrative pattern within these shaping books illustrates the female mentorial tradition characterized by Mitzi Myers, in which mothers or mother surrogates have authority as teachers and embody strength and rationality. Botany books in the familiar format featured mothers teaching botany to their children and using botany to teach broader cultural lessons. Writers of books in the familiar format promoted botany as a teaching tool and as part of a mother's responsibilities in early childhood education. The popularity of titles reprinted ten to fifteen times over a thirty-year period suggests that didactic introductory botany books served the purposes of various communities of interest among publishers, authors, book buyers, and readers.

Priscilla Wakefield

Priscilla Wakefield's *An Introduction to Botany, in a Series of Familiar Letters* (1796) was the first botany book written by a woman to provide a systematic introduction to the science. Shaped as the correspondence between two teenage sisters, it is addressed to children and "young persons" of indeterminate age—past infancy and early childhood but not yet into adulthood. In the frame established for the correspondence we learn that

AN

INTRODUCTION

TO

BOTANY,

IN

A SERIES OF

FAMILIAR LETTERS,

WITH ILLUSTRATIVE ENGRAVINGS.

BY PRISCILLA WAKEFIELD,

AUTHOR OF MENTAL IMPROVEMENT, LEISURE
HOURS, &c.

LONDON:

PRINTED FOR E. NEWBERRY, ST. PAUL'S CHURCH-
YARD; DARTON AND HARVEY, GRACECHURCH-
STREET; AND VERNOR AND HOOD,
BIRCHIN-LANE.

M,DCC,XCVI.

Priscilla Wakefield, *An Introduction to Botany, in a Series of Familiar Letters* (1796).

Felicia's mother and governess seek to raise her spirits during her sister's extended absence by teaching her botany. Botany is meant as a useful employment and amusement, to check her "depression of spirits" and give focus to her "morning and evening rambles." The governess (who will not permit her "to pass these hours in mere amusement") gives botanical lectures, and Felicia then communicates these to her sister. The letters provide an overview of Linnaeus's teachings for readers who are assiduous but not specialists. A series of plates illustrate the parts of plants, for example, the reproductive parts and the shapes of leaves and roots.

An Introduction to Botany supplies a clear and simple exposition of Linnaean classes, orders, and genera. One sister, after learning how to study the parts of a flower, describes each of the twenty-four Linnaean classes; their governess, her "attentive affectionate friend," has taught her about the contributions of Tournefort and Linnaeus to plant systematics, particularly about Linnaeus, whose "new system . . . may probably receive improvement from some future naturalist, [but] is never likely to be superseded." *An Introduction to Botany* incorporated botanical information found in handbooks such as James Edward Smith and James Sowerby's *English Botany*, and it illustrates Linnaean categories by reference to indigenous plants such as primrose and bindweed. Priscilla Wakefield also used Linnaean vocabulary to describe the essential characters of the elm flower: "The calyx has five clefts, and is coloured on the inside; it has no corolla, but the seed-vessel is an oval berry without pulp, containing only one seed, rather globular and a little compressed."[11] Just a few years before the appearance of *An Introduction to Botany*, Erasmus Darwin had gloried in Linnaean plant sexuality in "The Loves of the Plants," his celebratory poem about conjugal and illicit couplings in the Linnaean vegetable kingdom. By contrast to Darwin's lusty discourses of the body, Wakefield sidesteps or ignores the topic of plant sexuality. She uses Withering's bowdlerizing language of "chives" and "pointals" rather than "stamens" and "pistils" to designate the reproductive parts of plants, and, when discussing the taxonomic grouping of orchids, Gynandria, she claims ignorance of the sexualized coupling that the term denotes.

An Introduction to Botany, an expository book about Linnaean systematics, used botany to teach other matters as well. Enlightenment discourses of improvement led Priscilla Wakefield to itemize the general advantages of botany in this way in her preface: "[Botany] contributes to health of body and cheerfulness of disposition, by presenting an inducement to take air and exercise; it is adapted to the simplest capacity, and the objects of its investigation offer themselves without expence or difficulty, which renders them attainable to every rank in life." Although she inserted remarks about divine design into discussions—remarking, for example, that "the finger of the Divine Artist is visible in the most minute of his works"—her religious justifications for botanical study are perfunctory.[12] Her botany book belongs to informal education rather than to formal schoolwork, and she is principally interested in highlighting the general and educational benefits of plant study.

Wakefield particularly wanted to affirm the benefits of botany for girls.

Botany once belonged, she wrote, only to the "circle of the learned" who could read the principal texts in Latin; this "deterred many, particularly the female sex, from attempting to obtain the knowledge of a science, thus defended, as it were, from their approach." By contrast, accounts in English having made botany more accessible, "it now is considered as a necessary addition to an accomplished education." She proclaimed her aspiration concerning girls and botany this way in the book's preface: "May [botany] become a substitute for some of the trifling, not to say pernicious, objects, that too frequently occupy the leisure of young ladies of fashionable manners, and, by employing their faculties rationally, act as an antidote to levity and idleness." *An Introduction to Botany* therefore promoted botany in opposition to the fashionable life, and as part of a search for alternative models for girls. The author's strategy was to represent botany as a science that linked the cachet of a socially approved "accomplishment" to a philosophy of self-improvement. In the same way, she suggested botanical drawing as an allied area for work: accomplishment plus improvement. This is decorous science in the conduct book tradition. It distances girls from the aristocratic manners that many eighteenth-century educational reformers railed against and emphasized rational activities for middle-class girls instead.

An Introduction to Botany invites girls over the threshold into botany as an improving pursuit, but it does not challenge dominant ideologies about women, sexuality, politics, or religion. Its botanical teachings lack a visionary dimension and feature neither large romantic speculations nor grand taxonomic schemes that critique Linnaeus. The sedentary sister at home does not venture beyond the garden and adjacent fields, or beyond domestic ideology. Little can be read as subverting the existing social order. Nevertheless, Wakefield clearly drew botany into a female and juvenile orbit and recommended it for teaching girls good habits and healthy behavior. In contrast to many other writers of her day, she did not warn against female learnedness, nor did she elevate modesty, submission, and piety above all else.

Priscilla Wakefield (1751–1832) was a pragmatic reformer who wrote books that enlarged gender practices but did not radically challenge them. Her reference to botany as an "antidote to levity and idleness" shows an Enlightenment firmness of purpose characteristic of a versatile didactic writer, a liberal and a Quaker, who wrote out of a sense of service and also for money. She began publishing books when she was more than forty years old, and like other women in her day who turned to profes-

Portrait of Priscilla Wakefield (1751–1832) by T. C. W. Wageman, in *Lady's Monthly Museum, or Polite Repository of Amusement and Instruction*, August 1818.

sional authorship in their middle years, she took up the pen because of pressing family circumstances. Worries about her husband's lack of business acumen and a son's legal expenses led her into business arrangements with important juvenile publishing firms in London, including Elizabeth Newbery and the Quaker publisher William Darton.[13] Wakefield chose to write her introductory botany book, as well as a series of popular travelogues and natural history books, because they were new kinds of writing that she had reason to expect would be both popular and lucrative. Not a naturalist herself, nor a firsthand traveler to exotic places, she was a deliberate writer who saw her books as a way to earn money. Between 1794 and 1816 she wrote sixteen natural history books and travelogues for young readers, as well as *Reflections on the Present Condition of the Female Sex* (1798).[14] *An Introduction to Botany* appeared in the opening years of her career as a writer of popular science books, natural history miscellanies, moral tales, and travel books for "the rising generation." It was widely reprinted, came to be known as "Wakefield's Botany," and remained a staple of publishers' lists until 1841. Translated into French, it also appeared in three American editions. An appendix added to the tenth edition supplied an introduction to the natural system of plant classification.

The popularity of Wakefield's botany book suggests an alignment between her publisher, the intended reading public, and an author. Priscilla Wakefield identified a publishing niche for her book and aimed for a moderate price, few technical terms, and an easy familiar format; she considered *An Introduction to Botany* innovative in the way it adapted material for an audience of "children or young persons." In the 1780s and 1790s many books were written to teach adults about botany, but as she explained in the preface: "It appeared that everything hitherto published was too expensive, as well as too diffuse and scientific, for the purpose of teaching the elementary parts to children or young persons; it was therefore thought, that a book of moderate price, and divested as much as possible of technical terms, introduced in an easy familiar form, might be acceptable." When Rev. W. Bingley issued his *Practical Introduction to Botany* (1817), a small manual for amateurs, a favorable review declared that "it may with advantage be used as a Supplement to Miss Wakefield's Familiar Introduction to Botany."[15] Her book took on iconic status as well. In Sarah Atkins Wilson's *Botanical Rambles, Designed as an Early and Familiar Introduction to the Elegant and Pleasing Study of Botany* (1822), well-placed children take up botany as a topic of conversation, and the eldest sister often draws on Wakefield's Botany, as she calls it, for her botanical information. Bringing it to the parlor as her textual authority, she echoes the "good Mrs. Wakefield" about reasons for girls to study botany and praises their mother, who "wishes us to pursue this fascinating science, not merely as an amusement, but as a pleasing change for some of the trifling objects that too often occupy the time of many young ladies."[16]

Writing during the mid-1790s, Wakefield adopted an epistolary narrative to teach girls science as part of their general education. Her use of botanical correspondence adds another dimension to the catalog of fictional letter writing by women.[17] Her familiar format, with its domestic setting and familial cast of characters, authorizes female scientific curiosity. Her girls look at plants, are intellectually active, and teach one another in a sisterly relationship that includes shared intellectual enthusiasms. Wakefield made the letter a narrative vehicle for intellectual self-improvement in *An Introduction to Botany*, and her book models pleasure in knowledge. This may be the delight of the Quaker in the wonders of the natural

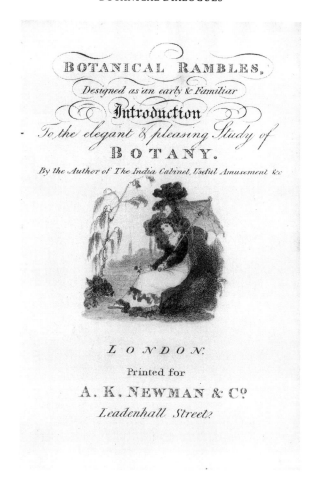

Sarah Atkins Wilson, *Botanical Rambles* (1822). (Osborne Collection of Early Children's Books, Toronto Public Library)

world. It is also a woman's delight in the pursuit of knowledge, as well as a woman writer's satisfaction in cultivating new markets for her literary and scientific work.

Sarah Fitton

Sarah Fitton's *Conversations on Botany* (1817) joined the roster of popular books teaching Linnaean botany at a time when the Linnaean method of plant identification was beginning to be supplanted. Organized as eighteen dialogues between a mother and her young son, this elementary textbook continued in print until a ninth edition in 1840. A reviewer in

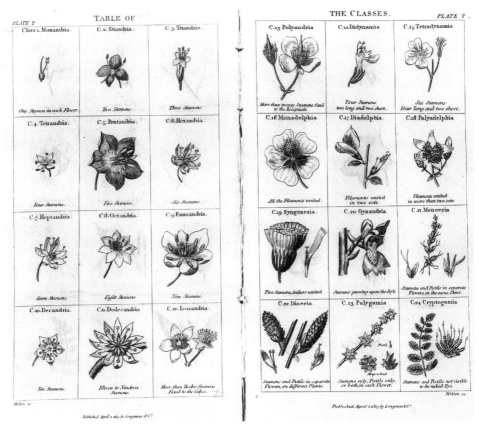

Sarah Fitton, *Conversations on Botany* (1817), plate 2, "Table of the Classes."

the *New British Lady's Magazine* commented that *Conversations on Botany* "manifests an advanced progress in the science, and instructs in a manner peculiarly attractive."[18] As Greg Myers has shown, the conventions and codes of pedagogical dialogues include figures of a simpleton and an author; typically someone wants to know something, or an adult has reasons for teaching about a topic. In the case of Sarah Fitton's book, curiosity supplies the opening when young Edward sees his mother examining a small yellow flower and asks, "What are you doing, Mama? . . . How do you examine flowers?" Mother and son walk together in garden and fields to look for examples of native English plants that match the Linnaean categories and the descriptions in William Withering's handbook, *A Botanical Arrangement of All the Vegetables Naturally Growing in Great Britain*. The conversational mode gives the appearance of inquiry, spontaneity, and exchange, but control remains in adult hands; fictional dialogues in science "suggest an idealized pedagogy initiated by learners but controlled in all its details by teachers."[19] In a conversation about poppies, for example, Edward's mother guides him into identifying the order as Monogynia because his poppy has only one pistil, then through the three genera that make up the order Monogynia to the genus of his poppy, based on the number of petals in the flower and the shape of the seed vessel. By the end of the conversation Mother has shown Edward how the Linnaean system has helped him identify his poppy as *Papaver rhoeas*, the common red poppy.

Conversations in Sarah Fitton's introductory book move from plant identification into discussing the uses of plants, particularly domestic and culinary applications (e.g., flax for linen, paper, and linseed oil). The mother teaches about the utility of plant knowledge, such as being able to distinguish between harmful plants and useful ones, and highlights the importance of botanical knowledge for physicians as well as for farmers. The book contains neither botanotheology nor natural theology—no suggestion that the orderliness of the vegetable kingdom demonstrates divine design. Sarah Fitton, instead, interpolates a story, "with no immediate relation to Botany" and drawn from a travel book, that "shows the value of knowledge and ingenuity in time of distress."[20] She completely sidesteps the issue of plant sexuality, making no mention of reproduction. In this book for children, young pupils, and parents, neither the mother nor the boy raises the subject; hence there is no room for the mother even to declare the unsuitability of the subject for her young pupil. Whereas some female authors flag their deliberate omission of "unsuitable language," here there is only silence.

Conversations on Botany is a dual-audience book that serves the pupil as well as the reading mother, who learns in turn how to teach her own children. The conversation is exemplary in substance; it educates the mother by giving her content and teaches her so that she can teach her children. It authorizes her knowledge. It also is exemplary in method, teaching women to be scientifically literate. The author of *Conversations on Botany* chose not to give the book a direct female focus; she and her publisher may have targeted an audience of mothers and small boys who were not yet with masters or sent off to school. Yet though the pupil may be male, his mentor is female. When the son inquires of his mother, "How did you find out so many curious things about plants?" she replies, "By reading different botanical works and books of travel."[21]

Conversations on Botany, published anonymously, was written by Sarah Fitton (ca. 1795–1870), assisted by her sister Elizabeth. The Fitton sisters lived in close family proximity to metropolitan science through their brother William, an Edinburgh-trained physician, man of science, writer on geology for scientific periodicals, and member of scientific societies. The Fitton family—a widowed mother and her daughters—followed William to Edinburgh, Northampton, and then to London, where he, having married advantageously, established himself as a gentleman scientist.[22] William Fitton regularly held Sunday evening scientific *conversazioni* for "a few scientific men," modeled on the gatherings of Sir Joseph Banks in Soho Square, and his circle included Robert Brown, the first clerk and librarian of the Linnean Society, later the "botanical legatee" to Sir Joseph Banks and in charge of the Banksian Department of the British Museum. Robert Brown, a man of high science, opposed popularizing directions in botany. When the Royal Botanic Gardens, Kew, and the British Museum were disputing with one another over control of important botanical collections, Brown showed his clear professionalizing orientation to botanical science; he "insisted that the activities of walking round a garden and of scholarly botany were quite distinct."[23]

Not surprisingly, given a professionalizing context of that kind, Sarah Fitton did not leap from *Conversations on Botany* into higher-level expository writing, nor did she write another botanical book for many decades. She may herself have been responsible for revisions and improvements to subsequent editions of her botany book. In later years she wrote short stories, as well as books for children, including *Conversations on Harmony*.[24] Her *The Four Seasons: A Short Account of the Structure of Plants* (1865) consists of lectures "Written for the Working Men's Institute in Paris." The genus *Fittonia* (Acanthaceae) was named in honor of Sarah and Elizabeth Fitton during that decade.

Sarah Fitton, like Priscilla Wakefield, chose the familiar format for her exposition of Linnaean botany. The mother in the narrative teaches informally in a domestic setting; her son is too young to have learned Latin yet but is old enough to move on from this book to use others without his mother's assistance. The publishing firm of Longman may well have been searching for an introductory botany book for young readers that would rival *Introduction to Botany*. In addition to using a conversational form rather than Wakefield's epistolary style, Sarah Fitton's botany book features a young boy as pupil rather than teenage girls and assigns teaching responsibility directly to a mother rather than a governess. Within the business of bookselling, these differences may have been important. But similarities between the two books in the familiar format are even more telling. Both Priscilla Wakefield and Sarah Fitton teach botany at home as part of general education and emphasize hands-on experiential knowledge of plants. Both authors give expertise to mothers and maternal-type teachers. The familiar is familial. Learners have personalities, and their situations have contexts. Science is to serve moral, spiritual, and intellectual purposes as well as being a source of pleasure.

Harriet Beaufort

During the opening decades of the nineteenth century, emphasis in popular botany books began shifting toward teaching about the structure, functions, and uses of plants. As botanists conducted experiments about nutrition, circulation of sap, and motion in seedlings, plant anatomy and physiology became topics of interest. Books and essays were published that introduced plant physiology to different levels of readers. Essays in the *Philosophical Transactions* during 1801–6, for example, described experiments about the ascent and descent of sap conducted by Thomas Andrew Knight, member of the Royal Society and president of the Horticultural Society. Publications such as James Edward Smith's *An Introduction to Physiological and Systematical Botany* (1807) helped move popular texts onto a new path by teaching about new findings in plant anatomy and physiology. Among the books that followed Smith's lead, Harriet Beaufort's *Dialogues on Botany* (1819) set out to teach plant physiology to "a younger class of readers." In forty-five dialogues about parts of plants and aspects of plant growth, the author discussed topics such as bulbs, roots, sap, fructification, modes of flowering, and the influence of climate on plant growth. The book teaches substantive botany with very few excursions into general topics. The author, clearly conversant with contemporary scientific literature, distinguishes her book from those that seek principally to teach Linnaean classification, "by which the young memory is loaded with systematic arrangements and technical terms, while the really instructive part of the science—the physiology of plants, and the progress of vegetation—are deferred till the ardour of the learner has subsided. This is to begin with the superstructure instead of the foundation."[25]

A serious-minded work, *Dialogues on Botany* also is a child-centered book with a distinctly progressive pedagogy. Harriet Beaufort explains that she chose the dialogue form as suited to children's taste, believing that children can identify with the characters. The narrative features three "very happy children, happy because obedient and industrious," who are taught about plants by a visiting maiden aunt who gives them "walking lectures" each morning. The book features a hands-on approach to studying plants and the tools of botanical study; the children dissect plants and work with microscopes to study the woody cells in the cuticle of roots, for example. There are no pictures because, the author avers in the preface, "children should be induced to study nature, rather than engraved representations." Since the focus of this book for young readers is plant physiology rather than Linnaean systematics, plant sexuality is not a central issue. In an aside, the narrator-teacher discusses stamens and pistils and offers to explain the bases for Linnaean classification to the children "if your mamma approves," but the subject is not raised again before the book ends.

Although the child-centered emphasis of *Dialogues on Botany* reflects practices in scientific narratives that had become common by 1819, this book has a more specific location in the circle of progressive educators of the period. Harriet (Henrietta) Beaufort (1778–1865) was related by marriage to Maria Edgeworth, novelist, pedagogical writer, and woman of letters. Her older sister Frances was the fourth wife of Maria Edgeworth's father, Richard Lovell Edgeworth. The Beaufort family, Huguenots based in Ireland, shared scientific and pedagogical interests with the Edgeworths as well. Harriet Beaufort was part of the Edgeworth family circle in Ireland and visited regularly at the Edgeworth home.[26]

Harriet Beaufort's botanical interests grew within her family. Her father, Reverend Daniel Beaufort, a parson-scholar and a founding member of the Royal Irish Academy, undertook many projects in science, technology, and topography. Plants, gardens, farming, astronomy, and mapmaking were among topics of interest to him, and the Beaufort children garnered the advantages of belonging to a scientific and liberal family. Her older brother Francis Beaufort went on to a long and illustrious career as hydrographer of the Admiralty, responsible for British marine chartmaking. Harriet herself "was wedded to her microscope and botanical books and specimens."[27] Dr. Beaufort brought his seventeen-year-old daughter a copy of Withering's *Botanical Arrangement* after a trip to London in 1795.

Harriet Beaufort's *Dialogues on Botany* appeared anonymously and was attributed at times to Maria Edgeworth. Her sister Louisa Beaufort wrote *Dialogues on Entomology* (1819), also issued anonymously. The two sisters, neither of whom married, kept house for their parents and other family members, including their brother Francis, at various times in the course of long lives. They are said to have written other volumes for children about natural history that appeared anonymously or under a nom de plume.[28] Their companion volumes on botany and entomology appeared in response to family financial uncertainties when the two sisters were in their early forties. Their father had relinquished his clerical position in 1818 and, burdened increasingly by debt in his later years, had mortgaged property that the children had inherited from wealthy maternal relatives. Both books were published by Maria Edgeworth's publisher Rowland Hunter; Edgeworth later helped to arrange state annuities for Mrs. Beaufort and her daughters when Dr. Beaufort died.

Harriet Beaufort's botany book illustrates the narrative pattern established for women's popular science writing by 1820. Placing science instruction within the orbit of children and their teachers, it not only includes girls but also features a female teacher. Her Enlightenment approach to botany recalls Priscilla Wakefield's book from twenty-five years before, as when Beaufort writes that "the more you converse on the subject of botany, the more you will be induced for your own sakes, as well as for the benefit of others, to furnish your minds with a store of this rational amusement, which will suit every period of your lives." She was well aware of other popularizations of science and referred to contemporary titles, including Wakefield's *Mental Improvement* (1794–97) and Maria Jacson's *Sketches of the Physiology of Vegetable Life* (1811). Expressing particular admiration for Jane Marcet's "delightful little book" *Conversations on Chemistry* (1806), she praised "the lady who wrote that elegant manual, for compressing so much instruction into so small a space, and for adapting the principles of an abstruse science to every body's comprehension."[29] Harriet Beaufort did not herself succeed, however, in accommodating her degree of learning to the narrative form of a popularizing botany book. *Dialogues on Botany* received mixed reviews. The *New Monthly Magazine* welcomed it as a "very useful elementary little book" and praised the use of dialogue for enlivening "the dryer part of the subject," but it declared the book "very defective" for not including illustrations, "which, in a professed elementary treatise, we humbly conceive to be an indispensable requisite." The *Monthly Review* compared it unfavorably with *Conversations on Botany* and criticized the author (referred to as male) for too much "attention to minutiae and too close an adherence to technical terms."[30]

Mary Roberts

Mary Roberts (1788–1864) often revisited the world of plants during a writing career that dated from the early 1820s to the early 1850s. She produced a steady stream of books principally for younger readers that combined nature, science, and religion. A versatile author, her publications include *The Conchologist's Companion* (1824), *The Seaside Companion, or Marine Natural History* (1835), *Sketches of the Animal and Vegetable Productions of America* (1839), and *Voices from the Woodlands, or History of Forest Trees, Ferns, Mosses, and Lichens* (1850). Her plant books diverge sharply from a secular Enlightenment culture of improvement and illustrate instead the new religiosity and evangelical piety of the nineteenth century. *Flowers of the Matin and Even Song, or Thoughts for Those Who Rise Early* (1845), for example, gives a "theme for meditation" at the times appointed for reading lessons in the Book of Common Prayer. Mary Roberts describes the evening primrose as a "nun-like flower" that "opens not til [the sun] is gone . . . like Charity, doing her good deeds in secret" and ends with thoughts about the Maker "who . . . provides for the wandering night ephemeron . . . when the sun withdraws his rays, [and] watches over even the minutest concerns of life."[31] Her approach to writing about nature was literary, appreciative, and religious rather than systematic and taxonomic.[32]

Mary Roberts's one introductory botany book, *Wonders of the Vegetable Kingdom Displayed* (1822), is saturated with the new religiosity. Like Harriet Beaufort's *Dialogues on Botany* that had appeared just a few years before, it teaches about topics in plant physiology, including perspiration, roots, plant nourishment, motion, and the dissemination of seeds. But Mary Roberts also extensively develops the theme of plants as moral and religious emblems. As the title leads one to expect, descriptions of plants such as the oriental poppy serve spiritual purposes: "Like the brilliant poppy [man] is not the flower of a day. The seeds of piety to God, and benevolence to man, are ripened in his bosom, destined to germinate and blossom in a richer soil, the garden of immortality."[33]

Wonders of the Vegetable Kingdom Displayed is written in the familiar format of letters but departs dramatically from the conventions cultivated by writers such as Priscilla Wakefield, Sarah Fitton, and Harriet Beaufort. Unlike earlier women botanical writers, Mary Roberts features neither a domestic setting nor a familial cast of characters and a maternal teacher. Additionally, she names her fictional correspondents Laelius,

THE

WONDERS

OF THE

VEGETABLE KINGDOM

Displayed:

IN A SERIES OF LETTERS.

BY THE AUTHOR OF

" SELECT FEMALE BIOGRAPHY," " THE CONCHOLOGIST'S
COMPANION," AND " LAST MOMENTS OF CHRISTIANS
AND INFIDELS."

" Not a tree,
A plant, a leaf, a blossom, but contains
A folio volume. We may read and read,
And read again, and still find something new,
Something to please, and something to instruct."
The Village Curate.

SECOND EDITION.

LONDON:
G. & W. B. WHITTAKER,
AVE-MARIA LANE.

1824.

Mary Roberts, *Wonders of the Vegetable Kingdom Displayed* (1822; 2d ed. 1824).
(Osborne Collection of Early Children's Books, Toronto Public Library)

Orontes, Timoclea, and Calista rather than the more British and everyday
Constance, Felicia, or Edward. Was Mary Roberts (or her publisher) in
search of a new narrative twist during the 1820s? It is possible that they
deliberately tried to distinguish their title from the familiar, familial, and
domestic pattern that had characterized popular botany texts since the
1790s. Alternatively, the change in formal features may represent a different set of political values during a decade of ferment and disruption.
Recently Nicola Watson argued that the epistolary form in fiction became
identified during the 1790s with sentiment, radicalism, and sexuality, and
that reaction to the French Revolution produced a backlash in narrative
form; more conservative writers "redirected" letters away from individu-
alism and into forms of social containment as a counter-revolutionary
tactic.[34] Organic metaphors in *Wonders of the Vegetable Kingdom* serve

conservative sociopolitical purposes; thus, we read that "society at large may be compared to a tree. The poor may be designated by the roots; the middle classes by the stem and branches; the dignified and noble . . . by the flower, leaves, and fruit. . . . To each an allotted duty is assigned." [35] Perhaps the reconfigured conventions represent Mary Roberts's counter-narrative to more progressive sentiments embodied in the epistolary format of Priscilla Wakefield's *Introduction to Botany* (then still in print). Compared with Wakefield's book, however, *Wonders of the Vegetable Kingdom* enjoyed only limited popularity. It was reprinted once and praised for "good sense, correct taste, and pious feeling." But one reviewer wished for "initials or plain English names" to describe the correspondents and quipped, "The taste for this sort of poetical names is quite gone out, and they are apt now to excite nausea." [36]

Mary Roberts promoted botany as a form of moral improvement and encouraged women to take up plant study. "It is much to be regretted," she wrote, "that individuals, who possess leisure for such interesting occupations, should not occasionally amuse themselves with experiments in vegetable physiology. To ladies especially, who are fond of horticultural pursuits, a new source of intellectual pleasure would be opened, by acquainting themselves with the causes of the frequent alterations in their favourite flowers." [37] She herself came to botany through family and religious heritage; her maternal grandfather was the learned seventeenth-century Quaker herbalist and botanist Thomas Lawson, and her father and brother contributed botanical material to the third edition of William Withering's botany book. She grew up in Gloucestershire, and years of residence in the country, coupled with her family background, Quaker nature-loving heritage, and study of writings by Withering, Cuvier, and Paley, led her to "an ardent love of natural history, . . . fostered, among the loveliest scenes in the west of England." [38] Her first publication, the anonymously issued *Select Female Biography, Comprising Memoirs of Eminent British Ladies* (1821), was inscribed "To the Ladies of Great Britain" as "examples of the illustrious dead [to] excite [readers] to higher attainments by the contemplation of merit." It features the piety and religious principles and practices of exemplary women from the seventeenth century on, such as the poet Elizabeth Rowe and the august poet and scholarly Bluestocking Elizabeth Carter. In later years Mary Roberts became known as "Sister Mary" because of the persona adopted in her book *Sister Mary's Tales in Natural History* (1834). She was a convert from Quakerism, perhaps leaving the Society of Friends at a point when many Quakers sought new forms of piety and public service in the evangelicalism of the established church. Mary Roberts did not marry, and she wrote in part to support herself and her widowed mother. Like other women in the Society of Friends and the evangelical movement, she used her scientific pen to serve both her religious ideals and her household economy.

Jane Marcet

Popular science books in the early nineteenth century introduced material to audiences who, because of sex, class, or level of education, were previously uninitiated into an area of science. *Conversations on Chemistry* (1806) is the paradigm of a book written for children and women that served as the equivalent of introductory textbooks for a general audience, having first inspired Michael Faraday with a love of science. It was the first of many popularizations of knowledge written by Jane Marcet (1769–1858), an energetic and prolific writer of introductory books, during a career that spanned the first half of the nineteenth century.[39] Although she wrote principally for young learners, her books served to extend education beyond childhood, to fill in the gaps, and to bring interested readers into the community of contemporary scientific knowers. At the end of the 1820s, Jane Marcet added botany to her repertory of popular expositions of science and issued *Conversations on Vegetable Physiology* (1829). This wide-ranging introduction teaches about the structure of roots, stems, and seeds and about plant functions such as nutrition and transpiration. It also includes applied topics such as transplanting, grafting, and plant diseases. *Conversations on Vegetable Physiology* is more learned and comprehensive in scope and level of information than other books by her predecessors among female botanical writers. As the author of an introductory chemistry book, Marcet was well versed in contemporary research (such as Priestly's work on oxygen) that was directly relevant to explaining plant structure and functions.

Jane Marcet blended substantive matter and narrative manner in an accessible way. Her narrative persona "Mrs. B." declared in the opening section of *Conversations on Vegetable Physiology*, "I am fully persuaded that no natural science is dry, unless it be drily treated." Like other women botanical writers, Marcet constructed her narrative for her readership. In this case she anticipated objections from female pupils who might find botany too hard or too uninteresting. The book opens by introducing Emily, a young woman wandering over beautiful mountains in Switzerland, who regrets her ignorance of botany. She would like to understand botany but lacks courage to make the attempt because, she explains, "there is such a drudgery of classification to encounter, before one can attain any proficiency to recompense one's labours." Her friend Caroline has no interest in botany as "a science of rules and names, not of ideas"; she is content "to gather a sweet-smelling nosegay of beautiful green monsters, as botanists denominate them, without troubling myself about their scientific names." Their friend and mentor Mrs. B. responds by describing herself as having earlier "entertained the same prejudices against botany." But she was "converted," she explains, by hearing lectures given in Geneva by Professor Augustin-Pyramus de Candolle that emphasized plant structure rather than classification and also included the application of botanical science to agriculture. From the vantage of a conversion experience, then, the narrative teacher offers an exposition of Candolle's examination of the vegetable kingdom.

To this end Jane Marcet employs the same conversational form she had honed in numerous previous books. The interlocutors give science a human face. The task for Mrs. B. is to convert two young women to botany. She is an alert pedagogue who adopts discourses and comparisons suitable for her audience and pauses when she judges that her pupils have had enough. Marcet, like her sister practitioners of the familiar format, folded extrascientific material into her exposition of botany. Thus, after Mrs. B. has taught her pupils about roots and plant nutrition, one girl declares that a study of plant structure and physiology is much more interesting than "the dry classification of flowers" because this approach "elevates the heart while it enlightens the mind, and bears more resemblance to lessons of morality and religion than to botany."[40]

Jane Haldimand Marcet, an Anglo-Swiss woman brought up in London who raised three children, had long-standing friendships with Mary Somerville, Maria Edgeworth, and Harriet Martineau. She and her husband Alexander, a physician and lecturer in chemistry at Guy's Hospital, were at the heart of a tightly interwoven mainstream social, literary, and scientific circle that included Mary and William Somerville, Humphry Davy, and W. H. Wollaston. They lived in Russell Square, London, held evening gatherings, attended lectures at the Royal Institution, and hosted gatherings of the Anglo-Swiss community. (Maria Edgeworth, visiting London from Ireland, reported on her time with Jane Marcet and "the famous learned Mrs. Somerville," saying that "all the best scientific and literary society drop into these two houses daily.")[41] After her husband's death in 1822, Jane Marcet divided her time between London and Geneva, a city that during the 1820s and 1830s enjoyed what a member of the Edgeworth family described as an "Augustan age" because of its mix of distinguished residents. Living in Restoration Geneva as a member of its liberal intellectual and academic circles, she joined with them in cultural activities that included amateur scientific study and pursuits.[42] Her botany book developed within the climate of Genevan polite science. Marcet had social and family connections to Augustin-Pyramus de Candolle, influential Swiss botanist and promoter of the natural system of plant classification, who gave formal courses on botany and zoology as well as "extraacademic" courses for "amateurs of both sexes" at the Academy in Geneva.[43] She attended the public lectures on botany and other scientific topics, and *Conversations on Vegetable Physiology* resulted, published when Marcet was sixty years old.

Jane Marcet mediated the botanical and scientific knowledge of her day by writing popularizing introductory books for young and female readers. She always published anonymously and, unlike her friend Mary Somerville, did not think of herself as "scientific." She positioned herself instead as a popularizer and a pedagogue. Harriet Martineau wrote that Marcet "always protested against being ranked with authors of originality, whether discoverers in science or thinkers in literature. She simply desired to be useful."[44] In the preface to *Conversations on Vegetable Physiology* Marcet claimed that all her knowledge of plant physiology came from Candolle's lectures, and that her contribution to the project was to arrange the material within a narrative form suitable for her intended audience. The skills required for the successful expository science writer to comprehend her material and shape it into a narrative are different in kind from research skills but should not be different in degree. The creation of knowledge and the dissemination of knowledge are equally important to science culture. Maria Edgeworth wrote about Jane Marcet: "I never knew any woman . . . who had so much *accurate* information and who can give it out in narration so clearly, so much for the pleasure and benefit of others without the least ostentation or mock humility. What she knows she knows without fear or hesitation and stops and tells you she knows no more whenever she is not certain."[45] Marcet's writing challenges us to think again about hierarchies of value that give only lukewarm consideration to "popular" science writing.

Although *Conversations on Vegetable Physiology* was less of a publishing success than Jane Marcet's other books about natural science, it was reprinted several times over the next decade. The *Magazine of Natural History*, trying hard to cultivate interest in plant study, was more directly enthusiastic, praising it as "a delightful book, written by a lady of high talent, on one of the most fascinating subjects which can engage the female pen. . . . We wish we knew what to say, in order to introduce this book into every family, either living in the country or having any prospect of ever walking in fields or gardens. We particularly recommend it to mothers and governesses; and we think it might be usefully introduced as a girl's school book in country schools, and as a prize-book."[46]

We have seen that one key feature of science books in the familiar format was that science learning took place at home. The familiar was familial and offered a representation of an ideal domestic life. Science learning was associated not so much with formal schoolwork as with one's other activities. Seen as a part of daily life and the general education of young people, science was also relational. Teachers and learners in this narrative form have personalities and stories and a setting in time and place. Epistolary and conversational narratives about science offered interpersonal connection and gave particular prominence to mothers. Inside science writing of this kind, a mother (or a mother substitute) teaches science to her children. Her interest in science is in phase with her other responsibilities, and science teaching is valued as part of good mothering. The maternal teacher, a figure of power and expertise for her children, is also an exemplar of female knowledge and intellectual authority for adult readers.

Inside the familiar science narratives, the mother is a complex figure, at once socializing and socialized. The representations "manage" women inside domestic life and assign them closely defined maternal duties. Managed mothers in turn, it can be argued, prepare women for life within patriarchal categories and within the constraints of patriarchy, albeit the new-style patriarchy of the Enlightenment.[47] But mothers and mother figures inside science books written in the familiar format can be interpreted another way—as giving women latitude to study science. Mothers there give science education a visible shaper, a personality and a voice. They provide a representation of female authority that is akin to the rise of female authority noted in domestic fiction.[48] A certain ideal type, self-possessed and self-controlled, she is rational mother rather than biological woman, moving beyond reproduction into rationality. The familiar format therefore offers at once a model of pedagogy, of motherhood, and of woman.

Pedagogical styles in books embody different cultural politics. Historically, the flowering of the familiar format brought women into science as teachers, learners, and educational writers. Books in the familiar format were one way for women writers to exercise their intellectual voices from the late eighteenth century into the early nineteenth. Those conventional forms suited women who wrote for money and were able to make new pedagogies work for them. (The conversational format, by illustrating good scientific conversation, also continued the tradition of training manuals in the art of conversation.) Intratextually and extratextually, then, popular botany books were a real resource for women. But by the early nineteenth century, however, and particularly by the 1830s, certain styles of popular scientific writing—the familiar format, poetic passages, generalist narratives with stories and digressions—became identified with women. Within botanical culture, popular books in the familiar format of letters and conversations were coded as "feminine" and tagged "nonscience." The broad history of women and science writing in nineteenth-century England records the separation and exclusion of women and the "feminine" from science and science writing as women were pushed to the margins of an increasingly masculinized science culture. Some women writers of botany books updated and diversified their narrative forms, but others cultivated careers on the margins.

FIVE

Three "Careers" in Botanical Writing

Having been requested by a gentleman, highly esteemed in the botanical world for the knowledge he has displayed in that science, to review the formation of seeds in general; to give a clear and concise picture of the growth of the embryo plant, from the first of its appearance in the seed vessel, to its shooting a perfect plant from the earth; to endeavour to prove the mistakes the variety of appellations have caused, as well as the misconceptions its extreme minuteness naturally occasions . . . honoured by such a request, I shall venture to begin my task.

Agnes Ibbetson, "On the Structure and Growth of Seeds," 1810

The liliaceous plants are six-petaled, or six-cleft; the style is triangular, and the capsule has three cells. Of the nineteen genera of this class and order, which grace the British Flora, several are remarkable for their delicate beauty; . . . Who is there that is not acquainted with the little drooping pearl that blooms beneath the snow, and seems as if it had fallen with it from the clouds? Who does not hail the pretty snowdrop?

Elizabeth Kent, "An Introductory View of the Linnean System of Plants," 1830

Throughout the Linnaean years, Flora's daughters were integral to the fabric of botanical culture in England as collectors, artists, students, and writers. No woman significantly reconfigured botanical knowledge by her published research, but a few women had "careers" as authors of books and essays that featured plants and botanical knowledge. For these women botany not only was an intellectual and economic resource, but was also central to their lives as writers. Maria Jacson, who published books between 1797 and 1816, hovered on the edge of the Enlightenment network that embraced Erasmus Darwin and provincial science culture in the Midlands; partly impelled by economic circumstances, she turned botanical interests into published botanical writing. In that same period Agnes Ibbetson, living deep in Devon and working as a self-described "recluse," was driven by a botanist's love of science to publish her findings about plant physiology and to seek substantial public recognition. Elizabeth Kent, at the heart of romantic London during the 1820s and 1830s, chose a literary rather than scientific route and parlayed a love of flowers and the literary discourses of romanticism into money and cultural engagement. Gender issues shaped how each of these three botanical writers positioned her work and how it was received. Maria Jacson benefited from an Enlightenment climate that afforded some intellectual elbowroom for women, but she had to negotiate the terrain between the Learned Lady and the Proper Lady. Agnes Ibbetson's botanical aspirations went beyond recreation; she chose a prominent public forum and wished to be received more seriously as a botanist than she in fact was. Romantic culture opened a space for Elizabeth Kent's linkage of plants and poetry but probably inhibited her intellectual work. The versatility and seriousness of these women is evident. Their careers as botanical writers depart from the popularizing modes of other women writers on botanical subjects, and their stories address issues of gender and authorship for women. Interleaving science culture and the intricacies of gender ideology, the careers of Maria Jacson, Agnes Ibbetson, and Elizabeth Kent illustrate ways botanical writing was a powerful resource for women, but one bounded by, and shaped by, literary conventions, gender ideology, and social constraints.

Maria Elizabeth Jacson and Enlightenment Botany

Maria Elizabeth Jacson (1755–1829) was a provincial gentlewoman who wrote three books about Linnaean botany and plant physiology and another about horticulture and garden design. Her books recapitulate developments in botanical culture during the later decades of the eighteenth century, for she came of intellectual age as a Linnaean and then turned her attention to plant physiology and horticultural applications of botanical knowledge. Less popular a writer than Priscilla Wakefield, Sarah Fitton, and Jane Marcet, she was better versed in her subject matter than many of that first generation of women science writers. Harriet Beaufort called her "an ingenious lady" in the preface to *Dialogues on Botany*, and praise of that kind echoed across reviews of her books.

Maria Jacson's first book, *Botanical Dialogues, between Hortensia and Her Four Children* (1797), taught the rudiments of Linnaean botany "for the use of schools." It is a compendious introduction to botany densely packed with information about Linnaean nomenclature and taxonomy. In the first section of the book the author discusses the reproductive parts of a plant, various modes of flowering, and classes and orders as two Linnaean divisions of plants. This "grammar" of botany in turn prepares students for "reading" Linnaeus's genera, the third division of the Linnaean system, so that they can identify plants properly. A second section of the book covers individual genera, such as the crocus and iris, along with ferns, mosses, and grasses. Jacson focuses on the reproductive parts of flowers without either embellishing or bowdlerizing her account. "The system of Linnaeus is called the sexual system of botany," she wrote, "because it is founded on observations, which seem to prove, that there are males and females in the vegetable world, as well as in the animal. The stamens are termed male, and the pistils female."[1]

Throughout the book the author argued strongly against popularized botanical accounts that oversimplify Linnaean ideas and language. She criticized "the pretty roundabout ways, which have been adopted to level the science to the capacity of ladies" and emphasized the value of using Linnaean botanical terminology and Latin botanical names for plants rather than common names or English generic names. Insisting that "scientific terms" are basic for conversing with botanists, Maria Jacson recommended in *Botanical Dialogues* that pupils learn botany by reading Linnaeus in a direct English translation, such as the translation of Linnaeus's *System of Vegetables* issued in 1783 by Erasmus Darwin and the

BOTANICAL DIALOGUES,

BETWEEN

HORTENSIA AND HER FOUR CHILDREN,

CHARLES, HARRIET, JULIETTE AND HENRY.

DESIGNED

FOR THE USE OF SCHOOLS,

BY A LADY.

" If we give our children nothing but an amusing employment, we
" lose the best half of our design; which is, at the same time
" that we amuse them, to exercise their understandings, and to
" accustom them to attention. Before we teach them to name
" what they see, let us begin by teaching them how to see.
" Suffer them not to think they know any thing of what is merely
" laid up in their memory."

ROUSSEAU'S LETTERS ON BOTANY.

LONDON:

PRINTED FOR J. JOHNSON, IN ST. PAUL'S CHURCH-YARD.

. . . .

1797.

Maria Jacson, *Botanical Dialogues, between Hortensia and Her Four Children* (1797).

Botanical Society of Lichfield. She is similarly forthright about the importance of studying plants directly, with one's own eyes, aided by magnifying glasses and microscopes. "Most of our botanical publications," Maria Jacson wrote, "are taken one from the other; and thus if an eminent botanist has in the course of his researches fallen into a mistake, the error has been propagated." She champions botanical works based on the rule "to *see for yourself.*" [2]

Botanical Dialogues is written in the familiar format of conversations at home. In sessions orchestrated by Hortensia, a botanically well versed mother and teacher, the children study plants as an amusement after finishing the day's formal lessons with their tutors and governess. Their botany lessons are informal, domestic, and site specific: "I have prepared this little room," Hortensia explains to her children, "which opens into my flower garden, for our study. Hither you may at any time come; and you will find books and glasses [that is, microscopes], and every thing that you may want." Hortensia guides the children along the path to substantive botanical knowledge and also makes botany a route to broader teachings. She is the earnest female mentor of the familiar format. Hortensia is convinced that botanical study will help her sons and daughters cultivate good mental habits and ensure that they do not "relapse into . . . indolent and desultory manners." Her teachings embody an Enlightenment recipe for botanical activity, according to which a good combination of amusement and instruction will make pupils agreeable, useful, and happy: "Every fresh acquisition of knowledge may add to our stock of happiness." *Botanical Dialogues* repeatedly highlights the ancillary mental and moral benefits of botanical study. Hortensia guides her children from specificities of Linnaean botany into topics pertinent to their general education. Thus discussions move from flaxseed to papermaking, from corn to Ceres, and from oats to straw and the invention of musical instruments. A child remarks on Hortensia's method: "You know, mamma, you always allow us to digress a little, when a new subject arises out of the old one, that we are studying; you say, it teaches us to think." [3]

The fictional family in Jacson's book represents a gendered and class-specific version of the utility of botany. The eldest son will become a gentleman of science in due course, and his mother accordingly recommends that he cultivate a taste for "useful and elegant studies" in order to be able to conduct philosophical experiments in a laboratory at home. The younger son, who "must exert his industry in a profession," will find botany useful if he chooses medicine. Hortensia's daughters will benefit from botanical education as an "amusing science" that teaches them to think. Hortensia also teaches her daughters, however, that "the education of the fingers" takes precedence over science because "the first point [is] to make ourselves useful in the small duties of life, which daily occur." [4]

The commendatory letter written by Sir Brooke Boothby and Erasmus Darwin that was published with *Botanical Dialogues* praises Maria Jacson for "so accurately explaining a difficult science in an easy and familiar manner, adapted to the capacities of those, for whom you professedly write." In fact, although the narrative form of *Botanical Dialogues* was meant to make the botanical teachings more accessible to parents and children, the book is overly long and relentlessly earnest. The familiar format seems not to have fitted Jacson, perhaps because she was more a botanist than an educator. In contrast to Priscilla Wakefield's *Introduction to Botany* (1796), reprinted many times over the next decades, Jacson's *Botanical Dialogues* received some positive reviews but never went past a first edition. On the one hand, Hortensia's lectures probably were too extended and complex for many young students, and on the other, the conversational format may have made the text inappropriate for better-versed readers. Darwin and Boothby suggested that Maria Jacson's book would suit adults too, as "a compleat elementary system for the instruction of those of more advanced life, who wish to enter upon this entertaining, though intricate study." [5]

Later Maria Jacson recast her first book into a form suitable for "persons of all ages." (As she explained: "The favorable reception given to my Botanical Dialogues . . . induced [me] to suppose, that the Work might be of more extensive utility, if divested of those parts peculiarly intended for the purposes of education, and altered to a form equally adapted to the use of grown persons as to children.") [6] *Botanical Lectures* (1804) is an exposition of Linnaean botany like *Botanical Dialogues*, but it elaborates on the importance of using proper botanical nomenclature and learning botany from the right sources. Maria Jacson situated this book as introductory to the translation of Linnaeus's *System of Vegetables* prepared by the Botanical Society of Lichfield (i.e., by Erasmus Darwin), "the only English work from which the pupil can become a Linnean, or universal Botanist." Contributing to contemporary debates about how to "English" Latin botanical terminology, Maria Jacson contrasted the exact translation of the Lichfield Linnaeus to others that only loosely adopt Linnaean terminology.

In *Botanical Lectures* Maria Jacson attempted to cross the divide by writing about botany for adult general readers. Her stylistic method was to replace the domestic conversation of *Botanical Dialogues* with expository prose without a narrative or a cast of characters. Instead of narrative aimed at a young audience, Maria Jacson cultivated a third-person format with the impersonal tone of the standardized science textbook. She intended the alterations to make the book usable beyond the schoolroom. The book exhibits the author's familiarity with secondary sources, as when she refers to recent botanical publications about mushrooms and grasses and discusses botanical debates about the classification of mosses. The confident tone and command of material in *Botanical Lectures* anticipates Mary Somerville's much more advanced scientific expositions of the 1830s. But here too, as in Maria Jacson's first book, there was a poor fit between the format and the intended audience. *Botanical Lectures* was pitched at too high a level for young readers and most general readers. At the same time, the authorial designation "by a Lady" probably meant the book would not have been selected for more specialist use. *Botanical Lectures* left little trace and was barely noted in periodical reviews.

With her third book, *Sketches of the Physiology of Vegetable Life* (1811), Maria Jacson joined a new generation of botanists—mainly Continental and to a lesser degree English—in turning away from taxonomy and pursuing questions about plant structure and functions. A reviewer of *Botanical Dialogues* had already commented on the author's "extensive acquaintance with many other branches of natural history" and invited her to employ her pen on "a subject intimately connected with that of the present volume, the physiology and economy of plants."[7] *Sketches of the Physiology of Vegetable Life* reflects wide reading in the expanding literature on this subject and shows Maria Jacson's familiarity with experiments conducted by English and Continental plant physiologists. She opened her discussion in this way:

> The taste for botany which has of late years so generally prevailed amongst all ranks and ages of society, and more peculiarly manifested itself in the younger part of the female sex, has long rendered me desirous to attempt to lead the more inquiring minds of those engaged in this interesting and rational pursuit, to a deeper investigation of the habits and properties of that division of organized nature denominated the vegetable kingdom, than is usually entered into by students of the technical side of the science.

Topics include analogies between plant and animal motion and sleep and reports on evidence that plants have a "faculty of sensation." Her intention throughout was to display evidence of the affinity between the vegetable kingdom and the animal kingdom. Through many examples, she opposes strictly mechanical explanations of plant motion and instead ascribes volition to plants. Writing about the hop and the honeysuckle, for example, she remarks that "plants of all kinds will make the strongest efforts to escape from darkness and shade, and to procure the cheering influence of the sun; this they will effect by bending, turning, and even twisting their stems, until they have disposed them in the manner best suited to receive the benefit of his vivifying rays."[8] In a similar anti-mechanistic vein, she reports on various ways that flowers "protect" and "nurture" young seeds.

Jacson addressed her book about plant physiology to a specific audience, "the young physiologist, who is only entering into the depths of this interesting science." She positions her discussion as an adjunct to several contemporary elementary treatises about plant physiology, citing in particular Erasmus Darwin's *Phytologia* (1801) and James Edward Smith's *An Introduction to Physiological and Systematical Botany* (1807). But *Sketches* is less a self-contained textbook than a book-length essay, one that aims to "excite rather than satisfy enquiry." Maria Jacson writes in a decidedly personal vein and describes her own observations and experiments. She reports experiments she conducted to ascertain the effect of light and heat on flowers of the spring crocus and recollects an experiment twenty years earlier on the material effect of the "honey" secreted by plants upon the reproductive parts. A series of drawings illustrate her findings about, for example, germination in a cocoa nut and the reproduction of several kinds of flower bulbs. Elsewhere in *Sketches*, Maria Jacson speculates about why trees drop their leaves in the autumn and suggests, in opposition to the prevailing belief that this phenomenon indicates disease, that it may result from a "natural process." Her belief in analogy between plants and animals leads her to the view that leaves "are the parents of the young buds which are found proceeding from their bosoms, by which their juices are absorbed, and which perish only when their offspring have attained their full period of growth."[9] *Sketches of the Physiology of Vegetable Life* has few counterparts among botanical writings by women. It is a botanical essay without narrative elements, familial personalities, or discussion about the wider benefits of plant study, and it meshes a first-person account of Jacson's own experiments with fluent remarks about contemporary plant physiology.

How did Maria Jacson come to know so much botany? Her story provides an instructive example of problems and possibilities in recuperative research. Her books appeared anonymously, the first two "by a Lady," the others "by the Authoress" of the first, identified in prefaces only by the initials M. E. J. No letters, diaries, or manuscripts are extant, and the historical record is scanty. Nevertheless, some information helps us reconstruct her career as a science writer and contextualize her work.[10] Maria Elizabeth Jacson's family was Midlands gentry, linked by marriage to lesser nobility. Her father and brothers were clergymen who held family livings in Derbyshire and Cheshire. Neither she nor her older sister Frances, a novelist, married, and both lived with their widowed father until he died in 1808. During that time Maria Jacson wrote her two Linnaean botany books. Sometime thereafter she and her sister moved into a handsome Tudor manor house in Somersal Herbert, a small village in Derbyshire (its population in 1801 was eighty-eight), where one brother had inherited a handsome Tudor manor house and another was church vicar. The sisters probably had some of the social status and responsibilities that belonged to clerical life. They also were linked to networks of provincial literary and scientific culture; in 1818, for example, Maria Edgeworth reported meeting them during a family visit to Lichfield. The novelist and pedagogical author wrote: "We have not yet seen any visitors since we came here and have paid only one visit to the Miss Jacksons [sic]. Miss Fanny you know is the author of Rhoda—Miss Maria Jackson the author of Dialogues on botany and a little book of advice about a gay garden." (She added, "I like the gay garden lady best at first sight but I will suspend my judgment prudently till I see more.") [11]

During this period, as we have seen, women studied botany as part of their own general education and learned to draw plants as part of an enlightened culture of accomplishments. Many botanically active women came into more sustained study by means of family networks that helped develop and focus their interests. In later years Maria Jacson referred to having "an hereditary liking" for flowers, which were "the most interesting amusement" of her youth.[12] Her clergyman-father may have encouraged his daughter to study botany for moral and spiritual reasons during her girlhood in a rural rectory. More likely, however, the pivotal figure in her botanical biography was Erasmus Darwin, whom she met through her cousin Sir Brooke Boothby. Boothby was a poet and botanist who belonged to Erasmus Darwin's Botanical Society of Lichfield and also to a local literary circle that included Darwin, Anna Seward, and Thomas Day. The artist Joseph Wright of Derby portrayed him in *Sir Brooke Boothby* (1781) reclining pensively in a verdant glade, holding a manuscript by Jean-Jacques Rousseau. Through family connections such as these, as well as shared interests in botany and natural history, Maria Jacson entered Erasmus Darwin's social and botanical orbit. Traces of her appear in his writings; thus we learn that Maria Jacson, "a lady who adds much botanical knowledge to many other elegant acquirements," gave him a drawing of a Venus's-flytrap in 1788 and reported observations to him in 1794 about the sagacity of a bird.[13] They probably had botanical conversations during the early 1790s about how to teach and promote Linnaean botany. The older man would have suggested books for her to read and would have expounded his strongly held views about how to translate Linnaean terminology.

In 1795 Maria Jacson submitted the manuscript of *Botanical Dialogues* to Boothby and Erasmus Darwin and invited their opinion of her work. Their joint commendation, written at a time when women botanical writers were neophytes on the publishing scene, no doubt helped her secure the publisher Joseph Johnson, who in turn capitalized on the power of Erasmus Darwin's name by placing Darwin and Boothby's letter at the head of the book. Erasmus Darwin later promoted *Botanical Dialogues* in his *Plan for the Conduct of Female Education*, his pedagogical tract for a progressive girls' boarding school that included botany in its curriculum. There he recommended *Botanical Dialogues* in preference to several earlier works: "An outline of Botany may be learnt from Lee's introduction to botany, and from the translations of the work of Linnaeus by a society at Lichfield [that is, his own]; to which might be added Curtis's botanical magazine. . . . But there is a new treatise introductory to botany call'd Botanic dialogues for the use of schools, well adapted to this purpose, written by M. E. Jacson, a lady well skill'd in botany."[14] Darwin, of course, was praising a book that echoed his own views on how to teach Linnaean botany.

Like a notably large number of the women botanical writers whose careers have been discussed thus far, Maria Jacson became a published author of botany books after she passed age forty. She may have written *Botanical Dialogues* from a pure love of botany or from beliefs about the mental and moral benefits of botany. She may have written it as a text-book for the school directed by Erasmus Darwin's daughters, the Parker sisters. Evidence suggests that she sought out publication of her work in part for financial reasons. Maria Jacson's father was in poor health in 1796 and made his will, "as it is of the Greatest Moment." He directed the disposition of household effects, including that Maria and Frances Jacson "shall take for their own use such part of the furniture of my house at Tarporley, of my linnen, plate, and china there, as they shall think necessary for furnishing such house as they shall hereafter choose to resettle in."[15] He stipulated a legacy for his daughters of fifteen hundred pounds apiece, an amount that would yield barely enough for two maiden ladies to keep a roof over their heads. (At the going interest rate of 5 percent, this amount would yield seventy-five pounds a year, a bit more than the stipend of a curate and less than the earnings of a skilled artisan.) Faced with such a prospect, a genteel spinster needed to plan how to supplement her limited inheritance. Popular science writing, as we have seen, was one route for women to take during that time, and Maria Jacson joined the cadre of women who wrote science books for children, women, and general readers.

Maria Jacson was part of the publishing stable of the progressive publisher Joseph Johnson, who also issued books by Erasmus Darwin and Mary Wollstonecraft. Both author and publisher could well have judged the moment advantageous in 1796–97 for an introductory botany book written by a woman and for children. The publisher Elizabeth Newbery issued Priscilla Wakefield's *Introduction to Botany* in 1796, and Joseph Johnson may have spotted a publishing trend in the making. The number of reviews of *Botanical Dialogues* suggests that the timing was right for a further introductory book. The *Monthly Review* gave considerable weight to the patronage of Darwin and Boothby and praised the book's language as "clear and elegant," adding that "the dryness of a mere explanation of the Linnaean system is relieved by the introduction of various amusing and interesting facts, and detailed observations." The reviewer continued: "It cannot, indeed, pretend to the sprightliness of Rousseau and his translator Martyn: but this is perhaps a necessary consequence of its closer adherence to the science."[16] Maria Edgeworth and her father tucked a few words of praise for Maria Jacson's work into their discussion on books for children in their pedagogical compendium *Practical Education*: "As this page was sent over to us for correction, we seize the opportunity of expressing our wish that 'Botanical Dialogues, by a Lady,' had come sooner to our hands; it contains much that we think peculiarly valuable."[17]

Jacson wrote her first botany book under the banner of late Enlightenment science culture, in the midst of a decade of debates about women's education and the status of women. At one point in the narrative in *Botanical Dialogues*, Hortensia suggests to her son that he might make botanical discoveries by studying ferns; the boy wonders why she offers this possibility to him and not to his sisters. She replies that women can be as informed as men in whatever area but should, as she puts it, avoid obtruding their knowledge upon the public. The world have agreed to condemn women to the exercise of their fingers, in preference to that of their heads; and a woman rarely does herself credit by coming forward as a literary character. The world improves, and consequently female education. Some years ago a lady was ashamed to spell with accuracy; happily the matter is now reversed, and the time will come, when it must be granted, that by improving our understandings, we enlarge our view of things in general; and thence are better qualified for the exercise of those domestic occupations, which we ought never to lose sight of, as our brightest ornament, when properly fulfilled. At this time, information in a woman, beyond a certain degree, distinguishes her above her companions, and like all other distinctions is liable to lead her into a vain display of what she hopes will gain her admiration. Hence she becomes ridiculous, and brings, what in itself might be a credit, into a disgrace whereas the disgrace ought to fall only on herself, and not stamp ridicule upon those of better understanding, who extend the advantages of their education to every occupation of life.[18]

These remarks evoke the figure of Mrs. Tansy in Charlotte Smith's *Rural Walks* (1795), a learned botanical woman who was criticized for vain display and tormenting "all the world with her knowledge" and was condemned to "piddle around" all alone in her botanical garden. Hortensia accepts a gendered dichotomy of head and hand and a separation of occupations: woman's sphere is domestic. Through the persona of Hortensia, Maria Jacson criticizes superficial female education but equally criticizes public female learnedness. She supports the liberal education of women but warns about the vanity that can result. In the long run, her argument goes, when more women are educated, the danger will diminish for individual well-educated women. Hortensia's comments in *Botanical Dialogues* may be Jacson's own voice or a more local strategy to forestall criticism of her learnedness. Writing at a time of political retrenchment, Jacson echoed exhortations about modesty and proper feminine practices that were widely found in conduct literature of the day. Even otherwise progressive pedagogues like Erasmus Darwin still maintained that "the female character should possess the mild and retiring virtues rather than the bold and dazzling ones."[19] Hortensia embodies knowledge, duty, and industry as a botanist and teacher, but she is always the Proper Lady rather than a Learned Lady.

Jacson's *Botanical Dialogues* and Darwin's *Plan for the Conduct of Female Education* both appeared in the year Mary Wollstonecraft died and within the anti-Jacobin climate that produced Richard Polwhele's attack on Wollstonecraft and female botanizing in his poem "The Unsex'd Females" one year later. Maria Jacson may well have sought to accommodate her science writing to an ideology of femininity—or at least phrased her work in terms of an appearance of that accommodation. Her book has no whiff of dissent, no glimpse of obvious political rhetoric. A reformist tendency shows only in relation to women's education, but even there she takes a long view. Compared with botanical-pedagogical books by the broad-minded and volatile Erasmus Darwin or the Quaker reformer Priscilla Wakefield, Maria Jacson's works camouflaged her political opinions.

By the time Jacson came to write *Sketches of the Physiology of Vegetable Life* (1811), the gusto of the knowledgeable, moral mother Hortensia had disappeared into a different narrative style. This book lacks the scope and conviction of the author's Linnaean expositions. There is no powerful mother to serve as the botanist's persona in a pedagogically secure context. Instead, discourses of modesty shape this book, and the author inserts many qualifiers about her "few partial experiments" and the value of her contributions. She repeatedly expresses the hope that her observations "may serve as a foundation for more extended experiments, by some ingenious students in the science of botany." Local considerations of health and family obligations may have shaped the direction of Jacson's botanical work, but a more global and more gendered explanation is warranted. A reviewer of *Sketches of the Physiology of Vegetable Life*, noting her obvious familiarity with the field and "the easy and unostentatious manner in which she communicates instruction," remarked that "the ingenuity of her reasoning and the diffidence of her manner are equally conspicuous."[20] This characterization pertains equally well to the later work of Maria Jacson and to the work of many women writers of her generation. Genre changes in Jacson's later books may show independence from literary convention, but the deference in *Sketches* suggests a writer who could not, or would not, leap fully into more substantial scientific writing.

For a woman writing about botany at that time, there were many reasons to mask "the ingenuity of her reasoning" by "the diffidence of her manner." Public visibility became problematic for women writers during the 1790s in the aftermath of the French Revolution as sociocultural boundaries narrowed for women and for women writers. Increasingly during this period, as Mary Poovey showed in studying the paradoxes of propriety among women writers who were Maria Jacson's contemporaries, ideas about the Proper Lady and femininity led women writers to self-expression characterized by indirection, accommodation, and self-effacement.[21]

Maria Jacson's books reflect a writer who sought elbowroom for herself and her women readers. After one experiment in the familiar format, Jacson departed from a family-based narrative and wrote *Botanical Lectures*, an exposition of Linnaean botany that dismissed gender-specific simplifications and made few compromises in its insistence on proper scientific vocabulary. Then, in *Sketches of the Physiology of Vegetable Life* she compiled a first-person account of her own botanical experiments and speculations. This, her most personal book, documents some of her own botanical experiments and speculations about plant physiology. Perhaps she found more freedom of voice in her later years, when she and her sister lived on their own after their father died.

Maria Jacson's last publication, *A Florist's Manual: Hints for the Construction of a Gay Flower-Garden* (1816), was a short book about garden design addressed to "sister Florists." It focuses on flower gardens, not on large vistas and greenswards in a landscape park, and teaches how to ensure "a succession of enamelled borders" through the spring and summer months by a mix of flowers, bulbs, and grasses. The aesthetic principle guiding the landscape design is the "well-blended quantity" of a "mingled flower garden" based on building blocks of color rather than "rarity and variety." At a moment in garden history when fashionable novices stocked their borders with rare species that produced little in the way of color, Jacson recommended color rather than rarity and common plants rather than exotics. Her book instructs novices about the layout of borders and parterres, "the disposal of a few shaped borders interspersed with turf."[22] A few engraved plans and a list of herbaceous plants help readers design their gardens and plan a sequence of flowers. The book also discusses methods for insect control and sets out instructions about cultivating bulbous plants such as the amaryllis.

Although the main emphasis of *The Florist's Manual*, a small practical book by a keen and experienced gardener, is aesthetic and horticultural rather than botanical, there too, as in earlier books, Maria Jacson endeavors to enlarge the mental terrain for women. At the end of the book she urges her readers to become "philosophical botanists" by exerting their intellects in "the study of vegetable existence" and recommends her own *Sketches of the Physiology of Vegetable Life* to that end. She understands well the gratification that can come from even a superficial contemplation of one's cultivated borders, but she hopes to "induce even a few of my sister Florists to exercise their intellect, or relieve their ennui" by inquiring about the causes of some things in their gardens (such as how bulbs sustain growth). Jacson wrote with an audience, and perhaps a publisher, in mind. Her final book appeared under the imprint of Henry Colburn, a publisher for the carriage trade, who also issued fashionable novels and popular travel books.[23] Perhaps because of Colburn's "puffing engine," his aggressive system of advertising books, *The Florist's Manual* was widely reviewed and reprinted several times. Interestingly, Jane Loudon, noted mid-Victorian author of gardening books, criticized the book in the 1850s as too erudite, high-flown, and ornate for novices.[24]

The excursus on women's education recited by Hortensia in *Botanical Dialogues* offers a perspective on Maria Jacson's own story as an unmarried woman, living at home as a clergyman's daughter, with domestic and social responsibilities. It helps interpret the direction (or indirection, or misdirection) of Jacson's career as a science writer, particularly why, despite her level of botanical knowledge, and despite one ambitious foray in *Botanical Lectures*, she remained within popular science writing directed to young people and women. Family networks gave this daughter of Flora access to influential individuals within provincial Enlightenment culture. Studious, industrious, with some independence of mind, Maria Jacson is a familiar female voice in her combination of learning and modesty. We can only surmise that she wrote her science books out of as much of an intellectual voice as she believed women should show, or dared to use in public. Both Jacson and her persona Hortensia navigated the fine line of contributing to the science education of girls, women, and general readers while not infringing on a provincial gentlewoman's schooled and well-honed sense of decorum. They inhabit the paradoxical late eighteenth-century terrain between the Learned Lady who tried to create more intellectual elbowroom for women and the Proper Lady who self-effacingly learned to accommodate herself to prevailing ideologies of femininity in her speaking and writing.

"For the Love of the Science": Agnes Ibbetson

During the late Enlightenment, it was congruent with gender ideology and a feminized botanical culture for women to collect plants, dry and

draw them, and write about botany in the form of popular and educational books for young and female readers. But what then of women
who wanted to study botany for reasons beyond amusement, recreation,
elegance, aesthetics, and pedagogy? Agnes Ibbetson (1757–1823) studied
plants and wrote "for the love of the science," as she put it. At a time
when botany was still largely Linnaean, tied to a natural history tradition
of description and classification, Agnes Ibbetson pursued serious research
into plant physiology. Her work bridged observational and experimental
methods. She dissected plants ("cut vegetables") over many years and relied on extensive use of microscopes. She believed that she had substantial
contributions to make, indeed that she had made discoveries about plant
physiology, and set out her findings in more than fifty essays published
in general science magazines; translations of some appeared in Swiss,
French, and Italian scientific periodicals.[25] When a plant was named in
her honor in 1810, the editor of *Curtis's Botanical Magazine* dedicated
Ibbetsonia genistoides ("spotted-flowered *Ibbetsonia*," now *Cyclopia genistoides*) to "MRS. AGNES IBBETSON, the author of several very ingenious
and instructive papers on vegetable physiology." [26]

The career of Agnes Ibbetson illustrates the vicissitudes of gendered
botany at that time. Contesting contemporary views about plant physiology, she was neither deferential nor apologetic in setting out her ideas
and disputed the findings of well-known botanists. Her published writings and drawings, and other unpublished materials, reveal a serious and
committed botanist who did not fit the feminized pattern for women during the early nineteenth century.[27] Botany was her all-embracing study,
and her methodological rigor and commitment to the science went beyond conventional limits. She also was a woman at odds with the male
power brokers of her field.

Agnes Ibbetson was among many researchers in early nineteenth-
century England, France, and Germany who debated topics in "vegetable
physiology," including seed formation, plant nutrition, and the structure
of roots, as they sought to explain plant processes. Whereas "animalists"
believed that plants were like animals, "experimentalists" were mecha-
nists who denied animal analogy.[28] Erasmus Darwin, for one, supported
animal analogy in his *Phytologia* (1800) and argued for an evolution-
ary gradation of being from vegetable into animal life. Agnes Ibbetson
was an "experimentalist." Dismissing the view that plants have sensitivity
and volition, she supported mechanical explanations of plant functions.
Plants, as she put it, are "machines governed wholly by light and moisture;

Signature of Agnes Ibbetson. (Courtesy of the Linnean Society of London)

Nº 1259.

Illustration of *Ibbetsonia genistoides* (now *Cyclopia genistoides*) (named in honor of Agnes Ibbetson), in *Curtis's Botanical Magazine* 31 (1810), plate 1259. (Courtesy of Hunt Institute for Botanical Documentation, Carnegie Mellon University, Pittsburgh) and dependent on these causes for motion."[29] She opposed the widely held belief that sap circulates in plants, as well as the idea that plants "perspire" and give off moisture and vapors through their leaves. Agnes Ibbetson had special interest in the structure and growth of seeds and believed that she had discovered a new part of the seed, the "nourishing vessels." One reiterated assertion was that "the Embryos of the Seeds are formed in the Root alone." She reports finding evidence from her experiments that seeds began as "a sort of gross powder," then became "very small balls, which afterwards entered the narrow passage of the middle root." Continuing her discussion of what now is called sapwood, she writes, "Here they generally stopped for a time, and then proceeding across the centre, entered the alburnum vessels in the stem, and mounted to the buds."[30] Throughout her writing she used terminology of the day that is now unfamiliar to us. She explained motion in plants in terms of the "spiral wire," vessels that are the "muscles" of plants. In her dissections she isolates "the line of life," "the impregnating vessel" that explains the vitality of each part of a plant and is found, as she puts it, "between the pith and the wood."

In terms of later botany, Agnes Ibbetson was mistaken in her theory about perspiration; modern accounts of plant physiology explain water on leaves differently, in terms of external and internal conditions that together produce transpiration. She also was mistaken in her theory that seeds form in the roots of plants, that the "whole progress of botanical vegetation" begins in the radicle, then passes into the stem. She was correct, however, in disputing the theory that sap circulates in plants as well as in her assertions about leaves as the lungs of plants. In terms of current botanical understanding, her accounts of "spiral wire" in plant tissue may refer to the distinctive spiral thickenings deposited on the walls of early-formed vessels (involved in water transport). But ultimate correctness was not, and is not, the issue in understanding Agnes Ibbetson's relation to the public community of botanists in her day. Within early nineteenth-century botanical culture, many botanists were disputing and jockeying for primacy over issues such as plant nutrition and plant motion. Personalities inevitably were part of this story, as were the parries and thrusts of contemporary debate among plant physiologists and other botanists. Like other botanists of her day who put forth theories, performed experiments, and wrote up their findings, Agnes Ibbetson wanted to be taken seriously for original experimental work. Her own material circumstances enabled her to dedicate time and attention to systematic scientific inquiry. But as an aging woman living far from a metropolis, she faced obstacles that were both external and internal. Her fierce determination to make her views known suggests the pattern of a self-taught scientist working without formal mentors and with limited opportunity for collegial contact.

Agnes Ibbetson (née Thomson) probably came into her botanical work through the popular and fashionable recreational route of polite science. Born into a merchant family in London, she may have been related to Thomas Thomson, chemist, teacher, and founder of the journal *Annals of Philosophy*. She attended a finishing school, and her early life, according to a contemporary, was "devoted to gaiety, frivolity, and dissipation, devoid of every idea of scientific or literary attainment." Her husband, a barrister, died in 1790 after a long illness, and she turned to "severer pursuits" in her middle years.[31] She had no children. From the late 1790s she lived on a comfortable annuity in the country, near Exeter in Devon. Journal entries, probably from December 1808, give glimpses of a quiet and studious life, in residence with a sister. She read Gibbon and did philanthropic work, "contriving for my Poor People, settling their Coals for the winter." She worked with a boy, perhaps a young nephew, reading to him in the evening. Her "usual morning employments" included mineralogy, experiments in galvanic electricity, and botany. Botany was her favorite pursuit, to which she devoted "13 out of 24 hours" over many years.[32]

Beginning with botany in the Linnaean mode, she first worked on grasses. Early portfolios contain over two hundred drawings of plants, along with Ibbetson's descriptive texts. The drawings are based on the Smith and Sowerby *English Botany*, and notes show that she grew some grasses herself, collected others in botanical rambles, and received dried specimens to assist her study. Her remarks show some independence, for she does not hesitate to criticize botanists such as Gerard, Stillingfleet, and Withering for insufficient information or mistaken identification. Ibbetson's working notes on grasses show that, although she had some interest in Linnaean description and taxonomy, she had more in plant dissection and microscopic research. She reports having dissected nearly one hundred specimens of wood bent grass (*Agrostis silvatica*) "at different times of its fructification, [and] . . . never yet could find a female pistil." In like manner, she refers to having examined two hundred specimens with a microscope to check a particular botanical feature.

By the early years of the new century, Agnes Ibbetson moved from Linnaean taxonomy into plant physiology. When she was in her fifties, she began submitting research reports to journals in which she described

---

A

JOURNAL

OF

NATURAL PHILOSOPHY, CHEMISTRY,

AND

THE ARTS.

SEPTEMBER, 1810.

ARTICLE I.

On the Structure and Growth of Seeds. In a Letter from
Mrs. AGNES IBBETSON.

To Mr. NICHOLSON.

SIR,

Having been requested by a gentleman, highly es- The author's
teemed in the botanical world for the knowledge he has dis- inducement to
played in that science, to review the formation of seeds in write on the subject.
general; to give a clear and concise picture of the growth
of the embryo plant, from the first of its appearance in the
seed vessel, to its shooting a perfect plant from the earth;
to endeavour to prove the mistakes the variety of appellations have caused, as well as the misconceptions its extreme
minuteness naturally occasions; and to show also in what
order the several parts appear, as physiologists have differed much in this respect: honoured by such a request, I
shall venture to begin my task, trusting in the great magnifying powers of the various excellent instruments we now
possess, and apologizing for venturing to contradict authors
so much superior to me in science, as in this matter the

VOL. XXVII. No. 121—SEPT. 1810.   B   eyes

Article by Agnes Ibbetson in Nicholson's *Journal of Natural Philosophy, Chemistry, and the Arts*, September 1810.

her experiments on plants and illustrated her findings by drawings based on microscopic observations. Nicholson's *Journal of Natural Philosophy, Chemistry, and the Arts* published thirty-three essays by her, often as the lead article, detailing her experiments on leaf buds, the roots and stems of trees, grafting, seed structure, and freshwater plants. She continued to report on her research findings in the *Philosophical Magazine*, the succes-sor to Nicholson's *Journal*, further clarifying her views on the interior of plants, plant nutrition, and the formation of seeds. Her twenty-two essays there include descriptions of dissections of the cuticle of leaves and air vessels in water plants as well as discussions about the chemistry of liquids and compounds in the atmosphere and in plants.

Ibbetson's botany was far removed from notions of female botanizing as elegant and delicate. Pursuing exactitude rather than elegance, she relied extensively on microscopes and dissection in her research. Microscopes were part of the eighteenth-century fashion of science in a culture enthusiastic about visual information, and women featured among those who used microscopes and telescopes recreationally, attended lecture demonstrations, bought instruments from various retail outlets, read explanatory books about microscopes, and carried "pocket glasses" with them to scrutinize plants during their morning rambles.[33] But Ibbetson went beyond recreational use of microscopes in her commitment to botanical study and her choice of instruments. She vociferously supported using them as a scientific tool and wrote about her own use of single-lens, solar, and lucernal microscopes. She credited her solar microscope with many of her serious botanical discoveries but gave first place to her single-lens microscope.[34]

Agnes Ibbetson also went well beyond recreational science in her commitment to dissecting plants. She was methodologically committed to regular and progressive dissection over a long period, as the basis for proper botanical research. For one experiment she dissected more than eighty wood samples. She also designed tools to aid her work. "I use almost as many different sorts as a surgeon," she wrote.[35] Her interest in developmental aspects of plant physiology led her to combine observational and experimental work. She wrote: "I may without exaggeration say, that I believe no one has ever dissected or watched plants with the unwearied diligence and patience that I have; taking up a fresh plant of the same kind every three days for nearly four years, following, watching the interior picture, and pursuing *each ingredient* from its first formation to its perfection, and hence to its destruction."[36]

Although she was an experimentalist committed to mechanical explanations of plant functions, Ibbetson personified nature as female and presented her work as explaining nature "herself." She described herself as "merely the transcriber of the information she [nature] gives." Her aim when she began dissecting plants was to "leave nature to write her fair history on my mind, and fix her various appearances in strong characters on my memory." Ibbetson was cautious about proposing grand systems. "It is an easy matter to establish a beautiful theory to captivate the imagination," she wrote, "but to understand every part of the formation of a plant, interior as well as exterior; to dissect and watch its various states and changes; to examine thoroughly how it passes into each, and what has been the general effect produced in the vegetable by such alterations; to collect by dissection and by culture its habits and powers—these are the requisites, and all this must be gained by examination and study, before we can at last form a theory founded on truth, and learn to know what a plant really is." She admits, however, that although she did not "set up to form a system," her thousands of dissections and thousands of sketches of parts of plants ultimately "created" a system for her. The works of nature, she wrote, are "perfect," with "beautiful simplicity"; and further: "I often throw down my pen and pencil in despair, ashamed of the folly of attempting to give an idea of works almost too great for man even to conceive. It is certainly true that we are more struck with the power of the Almighty, when we contemplate his small works, than in the view of his larger chefs d'oeuvre." [37] Ibbetson's accounts of her work habits recall Nobel Prize winner Barbara McClintock, a dedicated observer, who had "a feeling for the organism" and was committed to letting nature show itself to her.[38]

Agnes Ibbetson routinely criticized other botanists and hence needed to detail her methods as a way to defend her findings and establish the authority of her assertions. Philosophical botanists, she argued, systematized and made declarations about their findings but did not use their eyes enough. The botanist Jussieu, for example, created a "lovely fiction" when he arranged the vegetable kingdom by dividing plants into classes according to those with no cotyledons, those with one, and those with two. She shared his wish to find a natural method but regrets that his plan, "so simple and so beautiful," is false. Her evidence is that there are no plants without cotyledons and that the number of them in any plant varies greatly, often being more than the two he identifies. "Nor was Jussieu to blame; he is amply vindicated by the poorness of his magnifiers, by the knowledge of those times; though perhaps a little blinded by the beauty of his own conceptions. He examined with the powers he had; and, seeing all as others had done, he looked no farther." [39] She claimed that Thomas Andrew Knight, one of the premier gentleman botanists of the day, was likewise mistaken in his findings about the existence of sap vessels in the bark of trees because, she argued, he depended on chemical experiments rather than dissection.[40]

At a time when other women writers on botany fitted themselves strategically within gendered narrative forms, Agnes Ibbetson chose the scientific report as her medium. She submitted papers to Nicholson's *Journal* and the *Philosophical Magazine* as letters from a correspondent. Nicholson's *Journal*, begun in 1797, was a forum for both amateurs and experts and combined original articles with letters from correspondents and reprints of papers from other journals. (Volume 23, for example, included contributions by T. A. Knight, Humphry Davy, R. L. Edgeworth, and Arthur Young.) Ibbetson's letters did not differ in format from those written by male correspondents to these general science magazines, but her contributions take on a different hue in the context of her botanical sisters who wrote in the familiar epistolary format. For Priscilla Wakefield and Maria Jacson the epistolary form was a narrative convention in expository writing. As a popularizing narrative, the letter represented informality and relationship. By contrast, Ibbetson used it to communicate to the mainstream botanical community. Letters were not, of course, merely a conventional generic choice; they were communication, written from a domestic, private locus in the country to metropolitan periodicals and a community of the like-minded. But Agnes Ibbetson was the only woman to publish essays in Nicholson's *Journal* and later in the *Philosophical Magazine* and *Annals of Philosophy*.

Gender issues probably shaped how Ibbetson presented her work to these influential media of science and also how it was received. Well aware of her isolated position within botanical culture, she acknowledged at one point that her assertions about vegetable life would appear as "bold language, especially in a woman." [41] When she submitted her first two reports to Nicholson's *Journal*, she signed her letter "A. Ibbetson." The journal attributed authorship to "A. Ibbetson, Esq." and "Alexander Ibbetson," and she did not correct this until the conclusion of her third submission: "The mistake made by my directing my letters to be sent to Mr. I. has led you into an error. It is Mrs. Agnes Ibbetson, who has the honour of being your correspondent." [42] Thereafter she was "Mrs. Ibbetson." Early in the eighteenth century, men using female pseudonyms had contributed math problems to the annual *Ladies' Diary*. By the early nineteenth century the climate for female scientific contributions had chilled, and a scientifically-minded woman might easily think it advisable to mask herself on first entry into publication.

Did her findings carry less authority because they came from "Mrs. Ibbetson"? One contemporary commentator on "this ingenious lady's . . . very interesting observations and experiments" concluded that "the very high magnifying power she uses, aided by the warmth of her imagination, seems often to have led her into the regions of fancy." [43] Agnes Ibbetson accompanied her essays by intricate drawings based on her dissections and use of magnifiers. Details in some of her drawings have an aesthetic quality that led a current historian of scientific illustration to compare her work unfavorably with drawings done by others who used the same types of microscopes.[44] Yet there is a telling gap between criticism of Agnes Ibbetson's drawings as "fanciful" and her own accounts of her work. She wrote, for example, "I think I should not fear to stake my life on the drawings I have presented to the public: to be ready to lose it if any of them could be proved false, since every one has been drawn above 20 times."[45] Whereas some critics attacked her for the "warmth of her imagination," others criticized her for not generalizing more about her findings. Then, when she did put together a system, she was criticized for that too—told, as she put it, "that my dissections are admirable, but that my system is not admissible."[46] There was considerable debate in Ibbetson's day about accuracy in microscopy, about different types of instruments, and about the value of different degrees of magnification, and proponents of microscopes often were subject to criticism. But beyond all that, the reception of her work suggests how transgressive it was for a woman to conduct botanical experiments seriously, with commitment and absorption, to pursue exactitude rather than elegance, and to propose broad synthetic theories in science.

In 1814 John Bostock, a physician and botanist, encouraged her to bring together her findings and wrote to the newly knighted Sir James Edward Smith about his suggestion to Agnes Ibbetson, "a lady whose name you have heard, as an assiduous investigator of vegetable physiology," that she compile "a *connected view* of her experiments and discoveries on the vegetable economy." Bostock described her in his letter of introduction as "a most interesting character, full of energy and spirit, and completely devoted to her studies, which she pursues with unremitting ardor."[47] A few years earlier Smith had written *An Introduction to Physiological and Systematical Botany* (1807), a popularizing book about new discoveries in plant anatomy and physiology. Correspondence ensued between Ibbetson and Smith. She wanted him to recommend publication and to support having her findings included in the *Philosophical Transactions* of the Royal Society. The story, when pieced together from unpublished material over a two-year period, is painful to follow.

In May 1814 Agnes Ibbetson sent Smith her "Phytology," a forty-seven page letter explicating her "Philosophy of Botany" as well as folio-sized drawings of plant parts based on her dissections and microscopic studies and a further ten pages of notes. She explained that Dr. Bostock persuaded her "to draw up a short view of all the discoveries the Microscope enabled me to make" and wanted Smith "to consider the plan I have formed and whether it can be of use to the science I am so anxious to forward and improve." Ibbetson set out her ideas about the formation of flower buds, air vessels in plants, and the function of leaves and reviewed other key findings of her research. She roundly defended the value of microscopes for this kind of study and responded to criticisms of her drawings by invoking the high magnifying power that Leeuwenhoek and other notables used. "A Person who without the experience tries the solar, and *double microscope*, may find false lights, and deceptive shadows, but they all vanish before the being who, accustomed to the uses of the instrument, draws his conclusions not from one view but from a series of exactly pursued trials. . . . Experience and study are all that are wanted to perfect the view of such things. It is the person that *wants the practice necessary; not the Instrument which required improvement."* [48]

In August, having had no response from Smith, Ibbetson wrote again. She asked that he "suspend" his judgment and "not yet pronounce a sentence on me" until she sent him in a week or two "another paper which I hope will more completely finish the review of the Vegetable World." A letter dated December 12, 1814, asks if he received her "second set of troublesome papers" sent six weeks before. (That letter in the Smith Collection carries "ans'd" on the cover, dated December 21.) Smith eventually responded unfavorably. Although his actual reply is not extant, the few marginal comments on this "Phytology" manuscript show mild interest but stop halfway through the essay. In May 1816 Ibbetson wrote to Smith to express her "vexation":

> I beg you to believe I should not on any account have troubled you again on the subject had I not conceived my self authorized to do so by yr last kind letter. I am far from wishing to press you on any subject disagreeable to you. All I wished was a fair hearing. A Gentleman before whom I laid all the real specimens not in drawing but as vegetables, thought the proofs so unanswerable that they tempted me to trouble you once more. . . . As to publishing I do not suppose Rees would take the publication at his own without it was highly recommended and it is not at my age [that] people sacrifice fortune to vanity—I have little more than a comfortable annuity. — But my family are rich, and finding it greatly *patronized, they might*

have come forward. But this is over now—I can easily consign it to oblivion and the drawings and dissections of 16 years to the fire for I have no means of getting them received in the Philosophical Transactions where they would be well done. I have only to regret that I should have troubled you.

Instructions follow about how he can return her drawings via a bookseller in Exeter. She subsequently approached booksellers on her own, but they rebuffed her: "After consulting the first botanists . . . it was decided that no new facts were wanted."[49]

When Agnes Ibbetson sent the "Phytology" manuscript to Smith, she prefaced the substantive discussion by writing about herself. She describes herself as a "recluse" with no access to botanical gardens, books, or botanists: "From unassisted nature alone I can draw my means. . . . nothing but the consciousness of there being a series of absolute facts, supported by experimental proofs, . . . could give me courage to submit the whole to a Gentleman whose good opinion I value so highly." Her letter anticipates objections and has a predictably defensive edge. "I may say that my opinions have not been taken up lightly: many Friends around me have seen most of the experiments, and would willingly certify it." She justified her contributions by locating herself very carefully in relation to the work of other botanists, like Smith. Smith could have been more generous to someone whose views differed from his own. He, however—perhaps feeling challenged, in disagreement with her use of microscopes, her findings, or her drawings, or dismissing her as an older woman or an interloper in the field—clearly did not back her project.

It is illuminating to compare Smith's treatment of Ibbetson's manuscript with the respectful consideration that the botanist John Lindley gave in 1824 to a "curious" French theory of vegetable physiology by Aubert du Thouars, an author unknown in England. Thouars argued, among other things, that plants do not perspire. Lindley, discussing the theory in the *Philosophical Magazine*, explained that it is unnecessary to inquire whether the author's opinions "are founded in true or false reasoning. . . . They are at least the opinions of a man of high reputation as a botanist and philosopher; they have never been contradicted by any writer of talent, and from their peculiarity and ingenuity are certainly deserving of attention."[50] Agnes Ibbetson had discussed the same French botanist's work in Nicholson's *Journal* twelve years earlier and noted similarities between his findings and hers.[51] In one of her last essays for the *Philosophical*

Magazine, Ibbetson points out that assertions of hers that had been ridiculed some years before had been put forward in a recent *Transactions* of the Linnean Society and acknowledged to be correct: "How easy it is," she remarks, "to be prejudiced by the name of the presenter." [52]

In the wake of Smith's rejection, Agnes Ibbetson continued as a correspondent to the *Philosophical Magazine* and published in other general scientific forums. Like other plant physiologists of her day, she was interested in applied agricultural research, undertook experiments on soils, compost, and fermentation, and published papers on these topics. One paper, "On the Adapting of Plants to the Soil, and Not the Soil to the Plants," appeared in the proceedings of the Bath and West of England Society, a society that offered premiums for improvements in agriculture and planting. There she argued, based on her plant dissections and studies of how plants draw nutriments, that farmers would have a greater crop yield if they understood their soil and subsoil and chose plants that suited the particular ground. She also discussed how to use manures to build up soils. Her applied knowledge came, she explained, from access to a large farm and from sixteen years' experience growing potatoes; she also conducted experiments in soil improvement on a friend's estate, "treating his ground like a sick animal, and giving it plenty of wholesome nutritious food, but leaving to nature to bring up the proper grasses, nor counter-act her in any way." [53] Ibbetson published further reports on her experiments about soils in *Annals of Philosophy,* a general science journal with a special focus on chemistry that contained many articles about plants, particularly about the chemical components of nutrition, respiration, and transmutation. She was interested in studying decomposition in plants, especially how to produce effective "vegetable manure" by breaking down weeds into compost for soil enrichment.[54] Her essays record a series of experiments in soil science. For example, she buried weeds and plants in a deep trench and opened it every few months. Returning at regular intervals, she found confirmation of "the folly of digging up weeds and then burying them to make manure." She therefore recommends against burying weeds and grasses and recommends instead that farmers and gardeners burn them and then bury the ashes. She also experiments with using lime to bring boggy heath "into a perfect state of agricultural order and beauty in a short time." [55] Ibbetson's articles show her interest in bringing together agriculture and plant chemistry by the practical application of knowledge from plant physiology.

When Agnes Ibbetson died in 1823, she left completed but unpublished her "Botanical Treatise," the compilation of her research over many years. This two-hundred-page manuscript, written in the form of letters, reiterates her central beliefs about the mechanical nature of plants and about plant functions, including her conviction that plants do not "perspire." Among the topics are "On the Structure and Growth of Seeds," "On the Interior of the Stem of Plants," and "On the Formation of the Spiral Wire in Plants." The manuscript resounds with a blend of confidence and caution. Convinced about her findings, Ibbetson reiterates the methodological precision that led her to the conclusions she announces. Her confidence is based, she explains, on repeated experiments over a period of many years. Thus, she wrote:

> The constant habit of watching plants, at a very early hour in the morning, and very late at night, and examining them with a very powerful microscope, had almost convinced me that the Idea of the perspiration of plants was a false one: Still being acknowledged by all the first Botanists of Europe, it was necessary not to attempt to refute the mistake without the most absolute certainty of its being a deception of Nature, and without the contrary facts could be proved by the most undeniable evidence: This I now think I have obtained and that such proofs will appear in the next few letters as to deserve to be esteemed Truth.[56]

Agnes Ibbetson's treatise, a reprise of her botanical interests, is particularly poignant because it contains an "Address to the Public," probably written in the early 1820s and probably meant to be the preface to the manuscript. The document, as follows, holds a mirror to the status of women in botanical science, indeed in scientific culture, at that time.

> Aweful as it is, as it must be to a woman to present to the Public a work of science: The reflection that it is the result of near 16 years hard study can alone give me courage to offer it. The apparently daring plan of altering a science in all its parts from my own knowledge may revolt. I am truly sensible, and make my attempt appear too presumptuous. But with all humility, I may declare I never thought or imagined such a scheme, my whole Idea consisted in dissecting plants, and by never ceasing attention, care, and labour follow all the yearly changes both without and within the plant, in order to discover the course of nature thus hidden in her secret paths: This was undertaken for amusement: having no memory for

common botany, I chose a study more fitted to my abilities which I knew to be perseverance and the power of enduring excessive labour.

That I have succeeded even beyond my hopes, I have ventured to think, because when I began I did not flatter myself with ascertaining scarce a single truth: nor did I imagine that plants could be followed and developed in this manner. But it is certainly a much newer subject, a much easier one than was supposed. No hidden secrets—no occult miracles: The whole may (as I shall show) be followed up by the eye, with the help of a *single microscope*: The Hair alone requiring the compound and solar.

I present then the child of my old age to the public, and though the kind and favourable manner a mutilated part of this work was received has given me courage to complete it, yet I shall make use of no supplications, no excuses, no deprecatory speeches in favor of the work: The love I have for the science, instigated me to write it, and if the discoveries here made, will not be useful to the science; I do not even desire it may outlive its presentation: Not that I am indifferent to fame, the approbation of the public would gild with a glorious ray my declining years and even my death: But I wish only to obtain the suffrages of the public from deserving them. To that Public therefore I leave it well assured, if I merit its approbation, I shall receive it.

Agnes Ibbetson prepared this preface at the end of two full decades of active botanical work, during which she had reached out to the established botanists of her day. She had a palpable sense of grievance about the poor response to her work. She was at odds with the arbiters of botany in her day, and some botanists were at odds with her. Perhaps she would have received a better reception had she located her work more carefully in relation to current scholarship. But she was a woman, an older woman, and a latecomer to serious botany; she was an outsider who had no significant mentor, no buffers, no champion, no companionable strategist within the ranks of public botanical culture.[57] She would have benefited significantly from connection to a scientific society, but as a woman she had no access to membership in the Linnean Society, much less the Royal Society. Her sole institutional affiliation was to a primarily agricultural group, the Bath and West of England Society; she was named an honorary and corresponding member in 1814, and her paper "from the hand of a scientific Female Honorary Correspondent" was read at the annual meeting.

Agnes Ibbetson, experimentalist, was clearly disadvantaged by the lack of significant support and an active collegial scientific context. She did not fit the prevailing profile of botanical woman during the opening years of the nineteenth century. Although her essays brought her some public reputation, isolation and grievance were hallmarks of her career rather than the acceptance she longed for from a community of botanical researchers. Her story belongs to a narrative about women's entry and contributions to science culture, but also about a woman at odds with the male power brokers of her field and their unarticulated but evident gender norms.

A Romantic Flora: Elizabeth Kent

By the 1820s and 1830s, romantic writing as well as an anti-Linnaean direction in botany had diversified the languages of nature in books, essays, and poetry about plants. Elizabeth Kent wrote two books that were transitional between systematic accounts of plants and literary accounts. *Flora Domestica* (1823) and *Sylvan Sketches* (1825) integrate nature and narrative, science and art, and horticulture and romantic languages of nature in ways that signal changing directions in early nineteenth-century plant culture. Whereas Frances Rowden's expository *Poetical Introduction to the Study of Botany* (1801) used plants and Linnaean botanical information to teach manners and morals, Kent's books about flowers and trees are neither taxonomic nor didactic, nor do they belong to the genre of religious emblematic writing about plants. She recast botany in a romantic mode, combining systematic botanical knowledge with horticulture and a love of flowers. Interested in the specificities of plants, in flowers and trees as botanical specimens rather than principally as literary topics, she focuses on plants themselves rather than seeing them as a stimulus for reflection or imagination. Her writings, like those of Charlotte Smith, belong to the larger picture of women's writing during the cultural moment of English romanticism in being characterized more by closely observed particulars than by the visionary and transcendental fields of canonical male romantic poets.[58]

Elizabeth Kent lived and worked as a professional woman writer during the 1820s and 1830s within the personal network of romanticism. Her stepfather was the publisher Rowland Hunter, who succeeded to the firm

FLORA DOMESTICA,

OR

THE PORTABLE FLOWER-GARDEN;

WITH

DIRECTIONS FOR THE TREATMENT OF

PLANTS IN POTS;

AND

ILLUSTRATIONS FROM THE WORKS OF THE POETS.

―――――――

" How exquisitely sweet
This rich display of flowers,
This airy wild of fragrance
So lovely to the eye,
And to the sense so sweet."
ANDREINI'S ADAM.

―――――――

LONDON:
PRINTED FOR TAYLOR AND HESSEY,
93, FLEET-STREET,
AND 13, WATERLOO-PLACE, PALL-MALL.
1823.

Elizabeth Kent, *Flora Domestica, or The Portable Flower-Garden* (1823).

of the progressive bookseller Joseph Johnson and published Harriet Beaufort's *Dialogues on Botany* and titles by Maria Edgeworth. Her older sister Marianne married Leigh Hunt, romantic essayist, critic, and newspaper editor. Elizabeth Kent, who did not marry, participated in the Hunt social and family circle, in the verdant landscape of Hampstead Heath, London, where Keats and Shelley and other kindred romantic spirits came to visit. Letters by Mary Shelley offer glimpses of "Bessy" Kent visiting the Shelleys with Hunt nieces and nephews.[59] She shared her brother-in-law's literary interests and was Leigh Hunt's principal correspondent during the years when the Hunt family lived in Italy in the company of Shelley and Byron.[60] Writing for the most part anonymously, Kent produced and edited books and essays about plants. Her writing interests took root at home, and her first publication was *New Tales for Young Readers* (ca. 1818). Then, by her own account, "love of flowers and trees soon fixed my attention." When necessity forced her in later years to apply for assistance from the Royal Literary Fund, a charitable organization for writers, she supplied information about her background and the material conditions of her work that helps us reconstruct the career of a woman who tried to make her love of plants and flowers into an economic resource for herself.[61]

Elizabeth Kent's best-known book, *Flora Domestica, or The Portable Flower-Garden* (1823), describes in charming and informal style two hundred flowers, shrubs, and small trees that can be grown in pots and tubs, indoors on flower stands and outdoors on balconies. Container gardening had begun to spread in eighteenth-century England among town dwellers, who arranged their plants on windowsills and roof parapets, and the poet John Keats, living in suburban Hampstead, was among the early nineteenth-century enthusiasts of this kind of gardening.[62] Elizabeth Kent addressed her book to that practice. Her own sad experiences had shown her, she wrote, that even common plants like geraniums "often meet with an untimely death from the ignorance of their nurses," and hence she resolved, "in pity to these vegetable nurslings and their nurses, . . . to obtain and to communicate such information as should be requisite for the rearing and preserving a *portable garden* in pots."[63] Entries range alphabetically from agapanthus to zinnia and include advice on how to cultivate and propagate plants, and especially how to water properly. In addition to horticultural tips, *Flora Domestica* contains what Kent calls "biographies of plants"—anecdotes, poetic associations, and references to plants in folklore and mythology. The entry on anemones, for example, describes various foreign and native specimens of this flower, relates an anecdote about a miserly Parisian florist who imported an anemone from the East Indies and hoarded it for ten years, and then tells the story of jealous Flora's transforming the beautiful nymph Anemone into a flower.

Elizabeth Kent deliberately chose literary and horticultural languages for *Flora Domestica* as the way to capture readers for science by first interesting them in plants more generally. Reflecting years later on her early work, she described herself as "anxious to extend the study of [flowers and trees] among the numbers whom I saw deterred by the terms of science which met them at the threshold." She positioned herself in the same way as many other women popularizers whose writing aimed to help readers, especially young readers, cross into botanical study. "I sought first to excite an interest in flowers by the care to be bestowed on them and by treating of them familiarly hoped gradually to inspire courage to grapple with the scientific terms at first so formidable to young people." [64] To that end, she supplied only a small amount of botanical information in the book and deliberately veered away from a systematic approach to nature. Instead she emphasized plants for beauty and delight and focused on flowering plants rather than on ferns or foliage. Because "lovers of nature are most frequently admirers of beauty in any form," she also interpolated many poems into *Flora Domestica*. Verses from classical and contemporary poets illustrate her entries. These range from Virgil, Chaucer, and Petrarch to Wordsworth, Charlotte Smith, Keats, Shelley, John Clare, and Leigh Hunt himself, who helped her shape the book. *Flora Domestica* is, as a result, a sizable compilation of contemporary romantic verse.

Elizabeth Kent again combined plants and romantic verse in her next book, *Sylvan Sketches, or A Companion to the Park and Shrubbery* (1825). There she described eighty common hardy trees and shrubs—beech, bramble, maple, hawthorn, sumac—and explored historical and literary topics relating to trees. She wrote for a general readership and deliberately avoided botanical language in order "to give an unceremonious introduction of certain trees and shrubs." It is characteristic of her writing that descriptions are, as it were, rooted in particulars about trees. As in *Flora Domestica*, the narrative method begins with the plant and then moves out to cultural associations. Her entry on the plane tree, for example, describes its appearance, size, and leaves and then gives the history of its introduction into England during the sixteenth century. She begins an extended entry on the elm tree by surveying several types of *Ulmus* (common elm, American elm, dwarf elm) and their uses and then describes noteworthy elm specimens in England (e.g., one of Sir Francis Bacon's elms planted in Grays Inn Walks, London, in 1600, that was twelve feet in circumference when it fell in the 1720s). She also includes a long discussion about the threat from beetles that had already infested the elm trees in London parks during the 1820s. *Sylvan Sketches* contains many references to the uses of trees in other parts of the world as well; Kent cites explorers and travel writers such as Thunberg, Forster, von Humboldt, Brydone, Brooke, Clarke, Tooke, Thevenot, and Stedman and gives a rich cross-cultural account of the uses of trees. But as with *Flora Domestica*, this book is more about evoking feelings, pleasing the eye, and enlarging sensibility than about celebrating the narrow utility of plants. Elizabeth Kent intentionally moves beyond enumerating the uses of plants in her book because, she explains, "there is something beyond mere use, something beyond mere beauty, in their influence upon the human mind;— there is something in flowers and trees which excites our kindest sympathies, which soothes our keenest sorrows." Her own writing style is lyrical, evocative, visual, as when she described a grass (*Arrhenatherum avenaceum*) six feet high, with leaves two feet long and more than an inch wide, with its panicle of flowers gently drooping to one side, "so finely polished, that but for their green colour, we might think it was composed of silver oats. Yet it is not green, neither is it white, nor gold colour, nor purple, but it is a union of all these; it is the offspring of silver and gold, of the amethyst and the emerald." [65] Kent also illustrates the spiritual and aesthetic effects of trees by celebrating poets from Ovid to Spenser to contemporary romantics. Cowper, Coleridge, Shelley, Keats, Southey, and Hunt all make their appearance, as do prose writers such as Ann Radcliffe.

Members of the romantic circle recognized one of their own in Elizabeth Kent. Byron, Coleridge, Mary Shelley, and John Clare were all enthusiastic about her books. Coleridge, thanking his publisher for *Flora Domestica*, wrote: "Twenty times have I in the course of the last ten years felt and expressed the wish for such a work, as a Supplement to Withering. It is delightfully executed. . . . If I could write any thing that would be likely to further it's [*sic*] sale, I should consider it a sort of duty of gratitude for the delight I receive from Flowers the twelve months thro'." [66] John Clare, likewise the recipient of a copy from the publisher, declared himself "uncommonly pleased with *Flora Domestica*" and sent thanks to the anonymous author (who he assumed was male) "for the pleasure which the perusal of your book on plants has given me and for the use which it has given to society at large for every body who makes use of that mind which his creator has given him must feel delighted with the study of Botany." [67] Mary Shelley wrote that "Bessy's new book . . . forms with the other [*Flora Domestica*] a delightful prize for plant & flower worshippers—those favourites of God—which enjoy beauty unequaled and the tranquil pleasures of growth & life—bestowing incalculable pleasure and never giving or receiving pain." [68] Reviewers outside the romantic circle also were complimentary; one commentator called *Flora Domestica* "a convenient and agreeable guide" for "such ladies and gentlemen as beguile their leisure in tending their flower-stages, and are precluded from the accommodation of a garden" (but went on to criticize the book's lack of information about soil and then included a recipe for the right kind of compost for anemones).[69]

In 1828 Elizabeth Kent began writing reviews and essays for John Claudius Loudon's *Magazine of Natural History*, notably a series on Linnaean botany. Nine expository essays by "Miss Kent" guided readers through Linnaean plant classification and taught how to identify plants. One overarching intention of the series, congruent with the philosophy of the journal, was to persuade narrow utilitarians about the general value of botany. To that end, Kent cites examples of domestic culinary and medicinal uses of plants, such as the versatility of the elder tree (known for its narcotic scent and as a source for tea, wine, and pickled buds) and the efficacy of the roots of lily of the valley. (She cited without comment a recommendation from John Gerard's *Herball* (1597) that the fresh root of convallaria will help cure "any bruise, black or blue spots, gotten by fals [*sic*], or women's wilfulness in stumbling upon their hasty husbands' fists.")[70] But aiming to challenge and enlarge ideas about botanical utility, she discusses the uses of natural history not for its returns and profits but on a broader model of utility, one that includes beauty and poetry. "In a more liberal sense," she writes, "it may be observed that any thing which affords elegant and innocent recreation is of considerable utility."

Writing with a romantic sensibility, rhapsodizing about the beauties that botanical microscopes reveal in the colors and shapes of floral parts, Elizabeth Kent refuses to repudiate Linnaeus and botanical systematics. As part of her defense of Linnaeus, she explains that she used to agree with the position that "the study of botany detracts from our pleasure in the beauty of flowers" but now sees it differently. Names, she maintains, are just one branch of botany, and learning botanical systematics is akin to learning a foreign grammar so that one can use it for larger purposes. In this case the grammar of botany leads into the language of poetry. As she puts it: "These systematic niceties [are] a foreign grammar, which opens new stores of poetry hitherto unintelligible to us. The mystery that lies in the heart and first cause of everything, still remains the same, let us know as much as we can." Her essays, written at the end of a decade when Linnaean botany came under increasing attack, present the system of Linnaeus and the natural method of Jussieu as companion systems that serve different ends. Jussieu's natural method, she suggests, searches out connections and affinities among plants, and the artificial system of

Linnaeus acquaints us with individuals. "For a precise knowledge of the different species of plants, the system of Linnaeus is unrivalled."[71]

Elizabeth Kent's essays in the *Magazine of Natural History* promote botany as an area of study for girls, not with an eye to learnedness and specialized expertise, but as part of generalist knowledge. Kent believed that girls would do well to study plants. Nevertheless, she acknowledges that the Linnaean botanical system can be an impediment, not so much because of plant sexuality as because of the difficulties of the Latin language. Interestingly, Linnaean plant sexuality posed no obvious problem in her attempt to teach girls about botany; she mentions but does not dwell on descriptions of stamens and pistils as "husbands" and "wives." But the Latin terminology of Linnaean botany is another matter: "For a long period, indeed, science was carefully secured from the consideration of females by its locks of Latin; and, though the absurd prejudices against female education have, in a great measure, yielded to the progress of the times, yet that language, not being generally studied by ladies, still has power to scare them from an attempt, of which it leads them to overrate the difficulties." Elizabeth Kent therefore applauds translations of Linnaeus into English that have made botany more accessible: "The sex is considerably indebted to some naturalists of the present day, who have, as far as possible, anglicised the terms of science, and simplified the subjects of which they have treated."[72]

Elizabeth Kent's essays in the *Magazine of Natural History* in the late 1820s are an exercise in accessibility, but her assertions about girls and science are lukewarm when read in the context of the history of women and botanical culture, as though she is hesitant to push too far. Her tone differs from Enlightenment writers on this subject who considered botany an antidote to female frivolity. Moreover, her gendered thinking emerges when she compares the appropriateness of different areas of natural history. Girls can study plants, she explains, by reading books and looking at flower prints. Zoology, by contrast, is less suitable: a girl "will scarcely bring home in her delicate fingers a young hedge-hog, a hornet, or a bat, to compare with her books, and to ascertain their genus or their species; she will not dissect a monkey or a bear to study their anatomy."[73]

Much in Kent's introductory essay can be read as forestalling charges of learnedness leveled against the girls she sought to address in the series on botany. One wonders about the threat of learnedness as an impediment to her own writing career. She wrote in the late 1820s as though the 1790s never had been, as if books on natural history for girls and women had not been widely reprinted during the previous generation. Priscilla Wakefield's natural history miscellany *Mental Improvement* (1794–97) had been reprinted thirteen times by 1828, with its robust welcome for girls and their mothers into Enlightenment natural history study. Yet in a striking instance of changing views about girls' education and activities over a period of thirty years, Kent opened her discussion in the 1828 series in the *Magazine of Natural History* by stating that "girls are not only discouraged from the pursuit of natural history, but are very commonly forbidden it." In their younger years, she writes, instructors confound innocence with ignorance and keep them from books that contain information "which they would consider as objectionable, or excite enquiry which they would think it injudicious to satisfy." In later years, young ladies shrink in alarm from scientific terms and from Latin. This, she writes, has partly to do with "the dread of being branded with that fearful epithet—*blue!*"[74] Although her own writing contributed to the cultural project of promoting botany, and botany for girls and women, Elizabeth Kent, botanical writer, was herself probably hampered by gender norms in the 1820s.

Muted by ideological directions within her own immediate family and social circle, Kent likely was inhibited to some degree by fears of being labeled "blue." Some male romantic writers were particularly hostile to literary women and to intellectual women more generally. Keats wrote in a letter from 1817: "The world, and especially our England, has, within the last thirty years been vexed and teazed by a set of Devils, whom I detest so much . . . a set of Women, who having taken up a snack or Luncheon of Literary scraps, set themselves up for towers of Babel in Languages Sapphos in Poetry—Euclids in Geometry—and everything in nothing. Among such the name of Montague has been preeminent. . . . I had longed for some real feminine Modesty in these things."[75] Those who were not anti-blue often held clearly gendered notions about women and women's writing. When her brother-in-law Leigh Hunt reviewed *Flora Domestica* for his weekly newspaper the *Examiner*, he positioned women's plant work by describing Elizabeth Kent's book as "tying up its lady-like bunches with posies and ends of verses." He remarked how suitable it was that a woman should teach the care of flowers and intoned: "No pretension is made to anything great; but a great deal is done, which is very pretty and small."[76] Hunt was among the romantic writers who also appropriated nature for male study, colonizing for themselves sensibilities and literary topics, such as plants and gardens, that had previously been gendered feminine. Wanting to bring flowers and gardening within the masculine ken, he was intent on showing how many gentlemen loved flowers and gardens.[77] Although Leigh Hunt had helped Elizabeth Kent when she wrote her first plant book by proposing verses for inclusion in *Flora Domestica* and sending many editorial suggestions from Italy, he contributed far less to *Sylvan Sketches* and possibly did not know about this project until near its completion. There may be a note of bruised complaint in his remark that he had "never gathered from your letters that you were at all advanced in your tree book, much less that you had nearly done it." It is possible that Kent, gaining greater authorial confidence, kept him away from her second book.[78]

During the 1820s and 1830s, Elizabeth Kent was an active author living at the heart of publishing and literary London. *Flora Domestica* and *Sylvan Sketches*, successful publications for her, each went through several editions. Both books were issued by the firm of Taylor and Hessey, publishers of Keats and Coleridge. Leigh Hunt sent her ideas in 1824 for books on wildflowers and on female dress, books that, he said, would bring her fifty pounds each.[79] After publication of *Sylvan Sketches* she was asked to work on a book about British birds, and the poet John Clare was brought into the project. He was enthusiastic about *Flora Domestica*, told his publisher that he was "ready to join the Young Lady in writing the History of Birds," and even reported having been very busy "in watching the habits & coming of spring birds so as to [be] able to give Miss Kent an account of such as are not very well known in books."[80] Samuel Coleridge urged his publisher to commission her to write a book about wildflowers.[81] All these projects foundered, however, in the economic panic of the later 1820s, when the businesses of booksellers and bankers failed, eventually dragging her stepfather's business down as well. Kent therefore enlarged her vocational portfolio to include teaching botany as well as writing. She advertised in the *Times* in 1828 about giving young ladies instruction in the science of botany. The *Gardener's Magazine*, noting the advertisement, observed: "It is remarkable that this delightful science, which would seem to be more peculiarly a lady's study, has hitherto been almost entirely neglected by them; this may have proceeded, in some measure, from the difficulty in obtaining assistance; and Miss Kent is encouraged to hope that she may be the means of extending to others the pleasures and advantages she has herself derived from this innocent and interesting pursuit."[82] (She taught Linnaean botany for several years until she judged that the natural system had completely superseded the earlier taxonomy.)

It is not surprising that she established herself as a teacher of botany; she was fond of flowers in both their generality and their specificity.

After the publication of *Sylvan Sketches*, Elizabeth Kent continued to seek out opportunities to earn money through her writing. She projected a third volume about plants, this one on wildflowers. She wrote a chapter titled "The Florist" for *The Young Lady's Book* (1829), a compendium of topics, and published a collection of children's tales titled *New Tales for Children* (1831). She capitalized on her botanical knowledge by writing and editing books and papers. She prepared new editions of John Galpine's *Synoptical Compendium of British Botany* (1834) and Irving's *Botanical Catechism* (1835), correcting and updating each book in line with new developments in botanical science. Her preface to Galpine's book compared the Linnaean and Jussieuan systems and argued that they were complementary rather than rivals: "The Linnean System is the grammar of science," she wrote, "the Jussieuan is its literature." She would gladly have submitted other essays and reviews to the *Magazine of Natural History*, but her writing for that journal came to an end when the editor ceased paying contributors.

A litany of sorrows began when publishing opportunities dried up during the late 1830s. Elizabeth Kent became a governess for an attentive family and eventually received from them a modest legacy, which she used to establish a small residential school. By 1849 ill health, impaired sight, and business failure left her dependent on friends "of that class having more mind than money." She "still hoped to supply the means for [her] personal expenses by [her] own exertions" but described herself as "a stranger in the literary world. Instead of employment being offered me I mainly ask for it." [83] A publisher declined a manuscript that a friend had submitted on her behalf, about "a visit to the principal watering-places on the southern coast" with botanical information about plant habitats. In late December 1860 she was living with a nephew and his large family in Covent Garden, London, with no money for clothing, laundry, and medicine. The sad trajectory of Kent's later years resembles the profile of many other London-based literary spinsters during the early and mid-Victorian period who applied to the Royal Literary Fund for charitable assistance and received grants that averaged twenty to thirty pounds. [84]

Elizabeth Kent interjected a complaint into her introductory essay for the *Magazine of Natural History* about the status of women within botanical culture, specifically about their treatment in botanical gardens. In her day students of botany who wanted to move beyond indigenous plants into studying exotics could find many collections of exotics in private botanical gardens in the environs of contemporary London. But the proprietors of these "imported gardens" gave little assistance, Kent explained, to women who wanted to learn about plants. Most male visitors already had some acquaintance with botany. Among female visitors, many were interested in beauty rather than in botany; the great majority of "ladies . . . go merely as to an exhibition of beautiful plants, without knowing one from another, or caring to know them, unless, it may be, the coffee tree, and one or two others of popular note." But some women seek more information and a deeper opportunity for inquiry. What about them? Female visitors, she continues, are guided through these gardens too quickly, generally accompanied by "some ignorant lad, who is incompetent to reply to any question put to him, as to the plants, their names, countries, habits, &c." Kent did not mask the personal experiences that fueled her annoyance. "On some occasions, after asking several questions of the youth who attended our party, and finding that he could not answer them; that even the answers he did give were not correct; I desisted altogether from seeking the knowledge which I went purposely to obtain, and returned but little wiser. On one occasion, our party was accompanied by an able and experienced botanist; but there were older persons than ourselves, and gentlemen, in company, who, of course engrossed the whole of his time and attention; and those who were . . . the most interested in obtaining the information he might communicate, were precisely those who learned nothing from the visit but what their own eyes could teach them." [85]

Elizabeth Kent's account of gender politics shows botanical women in the 1820s caught between a "ladies'" culture of plants as beauty and a male culture of gentlemanly botany. The problem of women's relation to botanical gardens in the late 1820s was more far reaching than could be met by the short-term, local remedies suggested in her essay. Larger resonating issues in Elizabeth Kent's discussion include women's relation to learning and the status of women in the growing culture of expertise. Within romantic culture, the scientific woman was problematic. Within science culture, women were excluded from research-based botanical gardens. The status of botanical women became even more problematic over the next decades as cultural configurations around women and botany changed. Botany became increasingly masculinized. Women's serious conversations, marginalized as "polite," were interrupted within the budding science.

Defeminizing the Budding Science of Botany, 1830–1860

Flora's light-pictures are never repeated; her kaleidoscope is always turn-ing. Flowers are the universal moralists; not one but has its lesson, its sermon, or its song. . . . Faith and duty, and love and hope, and peace and gladness, smile on their dewy faces; fading in quiet hands, they speak of death; creeping over low green graves, they whisper of immortality. They are the emblems alike of feasting and mourning, of speech and silence, of sorrow and hope, of grief and love.

Chambers's Journal of Popular Literature, 1859

I have uniformly adopted the phraseology usually followed by DeCan-dolle, giving the name of *pinnae* to the primary divisions, and of *foliola* to the ultimate divisions. . . . I have also designated by *petiolus communis*, the *whole* of the stalk to which the pinnae are affixed, . . . and by *petio-lus partialis* I have meant the whole of the stalk to which the foliola are attached.

George Bentham, "Notes on *Mimoseae*, with a Short Synopsis of Species," 1842

In 1827 a biography of Linnaeus appeared that carried prescient messages about directions in botanical culture over the next thirty years. *A Sketch of the Life of Linnaeus*, written in the form of letters from a father to his son, is a heroic biography about the influential Swedish botanist and his travels in Lapland. A botanical "boys' own story," it is meant to correct a gender-coded representation of botany as only a "feminine" activity. The fifteen-year-old boy, away at school and being channeled toward medical training, needs to learn about plants but considers botany to be "not a subject suited to . . . the stronger powers of [his] sex"; he is "rather prone to think lightly of women's pursuits, as unsuited to the dignity of the manly character." The father endeavors to reclaim botany from the mark of exclusive association with women. He explains, picking up on the sex typing, that the boy has seen only his mother and sister involved with botany. The father used to be interested in botany, but "professional avocations . . . left [him] little leisure for it." Yet the boy must know about plants as part of his future medical training, and therefore his father wants him to learn botany from his sister "whenever you are home in the season of flowers."[1]

As a cultural text, Sarah Waring's biography of Linnaeus acknowledges the scientific engagement of young women and, like other books in the familiar format, enshrines the expertise of women in botany, at home and in relation to family. The boy's older sister, who lives at home, loves botany: "With a mind highly cultivated by the constant assiduity of her parents, and particularly of her amiable and intelligent mother, [she] was keenly alive to all the beauties of nature." But though she is well versed botanically, her knowledge is situated and circumscribed within the gendered parameters of her life. In this picture of the field, women give the boy elementary training at home, then hand him over to the men, who prepare him for public and adult responsibilities. The narrative embodies a model of male development in which growing up means going away from home and sister. On a sex-differentiation and separation model of maturity, hers is to teach and his is a way station to professional training and the public sphere. Within the book, the father, brother, and sister each embody gendered features of botanical culture that are a legacy to the early Victorian period. The author herself embodies complex gender ideologies of her day by writing in the male voice, ascribing authority to the father, and yet also assigning the daughter responsibility for informal scientific education.

Within botany, gender was one of the simmering cultural tensions during 1830–60, as academic botanists, writers, and protoprofessionals undertook to reshape plant study into "botanical science." Was botany an academic area or a popular one? How should it be taught? What were its languages? Who were its ideal students? These questions carried numerous consequences for different communities of interest, among whom were fellows of the Linnean Society, members of field clubs, popularizers, agricultural and horticultural botanists, publishers of botanical periodicals, students at mechanics' institutes, and women plant collectors.

During the early nineteenth century a new social map of science was being drawn in England. Specialist interest groups sprouted in London and in the provinces, and new disciplinary boundaries of natural knowledge were established. Gentlemen of science who were impatient with the generalist impulse of earlier scientific associations like the Royal Society found much to applaud in the emergence of subject-based groups. In London, as Geoffrey Cantor pointed out, "one could by the 1830s fill every weekday evening during the season with a lecture, a dinner or some other meeting of a scientific society."[2] Periodicals carried reports from scientific societies; in 1849 one could read accounts in the *Athenaeum* of meetings at the Astronomical, Chemical, Entomological, Geographical, Geological, Horticultural, Linnean, Microscopical, Statistical, and Zoological Societies. Gentlemen, political radicals, and "scientific activists" in the metropolis attended lectures and demonstrations at the Royal Institution, the London Mechanics' Institution, the National Gallery of Practical Science, the Royal Polytechnic Institution, and other public showcases that disseminated science. The scientific panorama of London was volatile and alive, "a flashpoint for controversy," peopled by widely contrasting groups and interests and with politics informing seemingly technical debates.[3] Audiences for popular science lectures ranged widely in age, station, religion, and politics. Michael Faraday gave Friday evening discourses at the Royal Institution for the social and intellectual elite of London and founded a juvenile lecture series there as well. The Royal Panopticon of Science and Art in London advertised in the *Literary Gazette* a "CLASS OF PRACTICAL CHYMISTRY, for medical students, gentlemen amateurs, or gentlemen wishing to investigate any particular branch of chymical science. A separate Class for Ladies, and a Juvenile Class in the morning."[4]

Practitioners, "gentlemen of science," hobbyists, writers, and public lecturers reflected in their various ways a growing bifurcation in science culture. As in other aspects of early nineteenth-century culture, there was a deepening divide between the general and the specialist, the popular and the academic, between the "high" science of gentlemen in metropolitan learned societies and the "low" science of practitioners who diffused scientific knowledge for practical use. By the 1830s, as Steven Shapin has put it, activities within scientific culture that were associated with politeness and gentility were rejected "in favor of a seriously impolite utilitarian culture," and speakers for utilitarian culture "articulated political demands for the professionalization of the scientific role and the state subvention of science—not because it fostered or was compatible with gentility but because it was materially useful to civil society."[5]

The professionalizing trend in botany took shape within the context of the Victorian romance of nature, for plants and flowers were among favorite objects in the Victorian cultural cabinet of natural history. Victorian naturalists went on excursions, gathered plants, studied specimens under microscopes, read journals, books, and picture books, and attended gatherings of local naturalists' groups and the perambulating annual meetings of the British Association for the Advancement of Science. An interest in plants and flowers satisfied diverse social, moral, religious, literary, and economic purposes at that time. Botanical avocations were congruent with Victorian values, including industriousness, a sense of awe and spiritual wonder, and the demands of evangelical ideology. The new religiosity within nineteenth-century culture shaped personal, social, economic, and political practices. Theological emphases on sin, retribution, conversion, and grace produced an evangelical fervor for practicing "vital religion," a puritanical piety that differed from utilitarian optimism and from the spirit of the Age of Improvement during the more rationalist Enlightenment.[6] Activities connected to nature came to carry a religious sanction and satisfied family and cultural mandates for "rational recreations." Pastimes with some moral or improving content gave the midcentury Victorian bourgeoisie, as Peter Bailey has shown, "a rationale which would relieve them of the need to apologize for their pleasures, yet still keep them within the bounds of moral fitness."[7]

In the midst of industrializing England, enthusiasts indulged their taste in horticulture, floriculture, and botany. Albums of pressed flowers sat in front parlors, breakfast rooms, and drawing rooms alongside cases filled with live and exotic specimens. Wax flowers, knitted flowers, paper

flowers, and shell flowers, floral fabric designs, tile designs, and naturalistic wallpapers were part of the visual field, and garden designs featured
floral borders, carpet bedding, and clock flowerbeds for the parterre in
place of the sweeping landscapes of trees favored during the previous
century. Flower shows and horticultural societies abounded in the metropolis and provinces, with shows and annual competitions. "In 1805,"
Ray Desmond points out, "only one horticultural society existed; by
1842 there were more than 200."[8] The workingmen's botanical societies
meeting in pubs in manufacturing areas of the industrial North continued an artisanal tradition dating back to Lancashire laborers in the
mid-eighteenth century.[9] Field clubs flourished in remote districts, larger
towns, and cities, drawing large numbers of men and women to day-
long outings for collectors, followed by an evening meal and discussion of
the day's findings; 550 attended one outing organized by the Manchester
Field Naturalists' Society.[10]

Victorians visited nature's floral curiosities in many social and geographical locations. In 1841 the Royal Gardens at Kew were made a public
botanical garden and soon became a place of pilgrimage under the direc-
torships of Sir William J. Hooker and his son Joseph D. Hooker, particularly when the Palm House was completed in 1848. The wealthy created
conservatories and heated greenhouses, stocked with exotics culled by
commercial collectors. The fern craze, or "pteridomania," blended aesthetics, science, fashion, and a certain type of Victorian compulsiveness.
From the 1840s through the 1860s, ferns were a fundamental feature of the
front parlor. A fernery was a desirable part of the home conservatory, and
Victorian parlors displayed their Wardian cases—glass plant cases much
like the bottle gardens or terrariums of the 1970s and 1980s. The repeal of
excise duties on glass in 1845, coupled with advances in sheet-glass manu-
facture, led to the grandiose designs of conservatories and the Crystal
Palace, and the Great Exhibition showed the potential of foliage under
glass. Enthusiasts collected ferns in the woods, bought tropical specimens
from nurserymen, dried and pressed ferns, and made spatter-work com-
positions (arranging fronds on paper and spraying them with india ink).
The rapacity of fern hunters who denuded the countryside in their search
for collectors' items provoked much comment.[11] Marine natural history,
including shells, seaweed, and cryptogams generally, also became popular,
fed by a cultural mix of fashion, medicine, aesthetics, and religion. When
Charles Kingsley, novelist and clergyman, railed in *Glaucus, or The Wonders of the Shore* (1855) against "the life-in-death in which thousands spend

the golden weeks of summer," he promoted the naturalist's activities instead: "Happy, truly, is the naturalist. He has no time for melancholy dreams. The earth becomes to him transparent; everywhere he sees significances." Botanizing at the shore during a family seaside holiday supplied that sanctioned mix of recreation and instruction.

Amid the various faces of Flora during the first half of the nineteenth century, then, different communities of interest formed, diverged, and sometimes overlapped. By the 1850s a new generation of middle-class "zealots," as David Allen termed them, moved into natural history culture in keen pursuit of fashion and respectability.[12] They expanded the social climate for natural history beyond the Quaker families, prominent dynasties, the Cambridge network, and Evangelicals descending from the Clapham sect who had formed the influential coteries in British natural history during the 1820s and 1830s. A self-sustaining cadre of academic and professional botanists was not in place in England until the 1870s, but distinctions were emerging and being established during the years 1830–60 between those with a more aesthetic, moral, and spiritual orientation to nature study and those with a more utilitarian or scientific approach. These different approaches to nature in turn shaped botanical education. Earlier decades had emphasized the social value and moral benefits of botany. By contrast, some Victorians shifted their focus from botany as general education to botany as technical instruction alone. Botanical culture increasingly offered contrasting assumptions about who was a "botanist" and how "he" would be taught. The botanist was distinguished from the botanophile, the scientific florist from the general reader, and love of botany from love of flowers.

John Lindley and "the Modern Science of Botany"

During 1830–60, the direction of botanical culture, like that of scientific culture more generally, was toward increasing stratification, and the "strictly scientific" botanical community began to specialize. John Lindley exemplifies the march toward a "modernized" and "scientific" botany. His career illustrates the vexed issue of systematics in early nineteenth-century botanical culture, and the growth of tension between popularizing and professionalizing impulses in botany. It also illustrates tension about the gender identity of botany and botanists. John Lindley (1799–1865) was the first professor of botany at London University (1829–60) and professor of botany to the Society of Apothecaries, and he also served the (Royal) Horticultural Society in various administrative capaci-

Portrait of John Lindley, 1848, from *Makers of British Botany*, edited by F. W. Oliver (1913). (Courtesy of Cambridge University Press)

ties for four decades. A ubiquitous and prodigiously energetic figure, he wrote voluminously about plants for various levels of audiences and produced educational books, technical monographs about orchids, a collaborative work on fossil flora, and botanical articles for the Society for the Diffusion of Useful Knowledge. Throughout his various areas of work, Lindley committed himself to putting botany on a new footing by cultivating interest in it as a science. When commissioned, for example, to survey the management of the royal gardens at Kew, Lindley recommended practices that would better satisfy his own belief that a botanical garden should be "a garden of science and instruction"; his report, submitted in 1838, led eventually to the Royal Botanic Gardens' being established as a national garden.[13]

Sir James E. Smith, founder and president of the Linnean Society, died in 1828. His death symbolized the end of an era in British botany, for though many popular writers continued to rely on the Linnaean sexual system as an accessible route to plant identification and arrangement, "scientific" botanists were turning at that point from taxonomy to physiology and morphology, and from the Linnaean sexual system to the natural system of classification. When John Lindley, then age twenty-nine, was appointed the first professor of botany at the newly established London University in the year of Smith's death, he set out to depose the Linnaean system and establish the natural system of botanical knowledge in its place. This system, promulgated by Continental botanists from Geneva and Paris, focused on morphological features of plants and natural affinities among plant groups. Augustin-Pyramus de Candolle, professor of botany in Geneva, following Antoine-Laurent de Jussieu, classified plants based on the totality of all organs rather than one fixed and discrete characteristic. Whereas Linnaeus used physiological characters and counted parts of the reproductive structure of plants, Candolle developed the concept of plant symmetry, accentuated morphological features, and focused on the positional relations of organs. (It was Candolle who grouped plants into monocotyledons and dicotyledons, the basic vocabulary still in use.) Robert Brown, his British contemporary and the most eminent British botanist of the early nineteenth century, using Jussieu's natural method of classification rather than Linnaean systematics, developed theories about floral morphology and natural affinities among plant groups, and his publications were influential among specialists in promoting the new method.[14]

When John Lindley delivered his inaugural lecture as professor of botany on Thursday, April 30, 1829, he aligned himself with Continental theories and roundly rang an anti-Linnaean theme. He argued that the "science" of botany should concern plant structure rather than identification—should be botany according to the newer Continental mode rather than on stale and static Linnaean lines. Linnaeus was a figure of historical importance, "a person exactly adapted to the state of science of the time in which he lived," but "a more advanced state of science" called for another system. Lindley acknowledged that the Linnaean system had the merit of simplicity, but he considered it to be superficial knowledge: "The principles of Linnaean classification produced the mischief of rendering Botany a mere science of names, than which nothing more useless can be well conceived." Instead, Lindley called for using a system based on plant structure: "Let the vegetable world be studied in all its forms and bearings, and it will be found that certain plants agree with each other in their anatomical condition, in the venation of their leaves, in the structure of their flowers, the position of their stamens, in the degree of development of their organs of reproduction, in the internal structure of those organs, in their mode of germination, and finally in their chemical and medicinal properties."[15]

In opposition to Linnaeus's sexual system of plant classification — theories that, though not British in origin, had become naturalized in Britain, institutionalized in the founding of the Linnean Society and effectively enshrined by the presence of Linnaeus's own herbarium in London — Lindley's call for a new system advocated a series of Continental theories about plant taxonomy. This was more than a botanist's decision about the shortcomings of a particular technical taxonomy. Indeed, Lindley's advocacy of Continental theories in his inaugural lecture echoed the pro-Continental, reformist attitudes of the new London University and reads like a profession of faith from within a broader sociopolitical credo. London University, founded in 1826, represented all that was new and radical and commercial in imperial London. This meant being opposed to the Tory-Anglican alliance and committed instead to reform and to training professionals for a new world of knowledge, including applied science. Science was one of the professional tools for the enlightened and secular reformers of the "godless college," as the nondenominational London University was called. The Benthamites of the new university "repudiated the aristocratic ideal of polite knowledge as an embellishment to a gentleman's education and sought to create a serious body of knowledge useful to the reformers in government and the professions. . . . [they aimed to] turn . . . middle-class students into a professional elite, a new middle management."[16] Lindley's botany belongs to that larger agenda, and Lindley himself, described by a biographer as "a first class fighting man,"[17] represents the emergence of a new generation. These new orientations led particularly to repudiating botany as polite knowledge and to embodying it rhetorically as "feminine."

John Lindley's inaugural lecture was his call to arms for a new botany for a new age. He rejected Linnaean systematics and botany as taxonomy and elevated the natural system and morphology instead. In so doing, he also rejected Linnaean botany in its social location in England. He declared his intention to "redeem one of the most interesting departments of Natural History from the obloquy which has become attached to it in this country."[18] The "obloquy" of botany in his day, as his language makes apparent, traces to connections between women and the polite culture of accomplishments. He put it like this in his lecture: "It has been very much the fashion of late years, in this country, to undervalue the importance of this science, and to consider it an amusement for ladies rather than an occupation for the serious thoughts of man."[19] In Lindley's formulation the Linnaean system was not only polite knowledge, but polite knowledge for women—"amusement for ladies." By contrast, Lindley's botany will be worthy of the attention of "men of enlightened minds." His chosen mandate, in his teaching and other botanical work, was to rescue botany for science, separating it from the realm of politeness and accomplishments that for decades had linked botany to women. His rejection of Linnaean botany is a rejection of polite botany in favor of utilitarian botany, and Lindley can be seen as wanting to shape a new man, a new kind of botanist, a scientific expert, fashioning a new identity for the scientific practitioner. Professionalization of botany meant its masculinization as well.

John Lindley's distinction between "science" and "polite accomplishment" publicly inaugurated a campaign in early Victorian England to defeminize botany by defining a scientific botany for gentlemen.[20] His bifurcated identification of polite botany with women and botanical science with men signals the direction botanical culture took during the succeeding decades.

The history of many disciplines duplicates the pattern found in nineteenth-century botanical culture: communities demarcating disciplinary practices, hiving off professional from amateur pursuits, distinguishing hierarchies of practice, and articulating appropriate levels of discourse.[21] The exclusionary practices of self-defining elites are a powerful and poignant part of the history of women and botanical culture. In Victorian science culture, two competing elites struggled for authority, as secular middle-class scientific naturalists and aristocratic Tory Anglican clergy and parson-naturalists contested directly and indirectly.[22] Women were one clear casualty in this struggle between science and religion. With gender as an analytic tool for recontextualizing the Victorian study of nature, one can interpret the nineteenth-century emergence of professionalized discourses in terms of a masculine "culture of experts" taking away authority that earlier had been invested in women's feelings and experiences. The "sciences of man," Anita Levy has argued, replaced the "sciences of woman" in domains as various as psychology, anthropology, and fiction, and disciplines gendered their cultural information and discredited certain cultural values.[23] In like manner, inserting a gender dimension specifically into the history of the culture of botany, we can read the professionalizing culture of botanists as endeavoring to remove its practices from the world of domestic, family, and personal relations—in short, from the realm of women and domains and practices marked "feminine."

Changing Narratives of Nature

By the mid-1840s the language of flowers and the language of botany diverged, and literary botany and scientific botany became distinct discourses. When George Bentham, a botanist of significant reputation, wrote about the mimosa (see the epigraph to this chapter), his account differed markedly from literary renderings and from those by Erasmus Darwin and Frances Rowden fifty years earlier. Darwin and Rowden, it will be recalled, had both versified about the mimosa in their own ways, personifying it as "chaste" and shy, a "tender plant" of "Sweet Sensibility." Of course formal botanical descriptions of plants and literary accounts are rhetorically separate; nevertheless, Bentham's technical discourse illustrates the increasingly specialized language of mid-nineteenth-century botany.

The rich soil of Victorian flower culture nourished abundant discourses of nature, including technical languages, floriography, nature poetry, and essays. The "language of flowers" once again became fashionable in England, and writers in this mode treated flowers as a poetic language, a constructed knowledge system with a universal code of meaning. The language of flowers, perhaps taking its European origin from Lady Mary Wortley Montagu, came to England via France, where a tradition of French sentimental flower books had taken shape during 1790–1820, culminating in Charlotte de Latour's influential *Le langage des fleurs* (1819). In floral alphabets and flower language dictionaries, writers attached sentiments to individual plants and developed a floral vocabulary for talking about emotions. In England, books on the language of flowers had a wide readership. Many interpreters studied flowers for moral meaning and spiritual enlightenment and, reviving seventeenth-century emblem books, read flowers as rich in Christian (and especially evangelical) application. "Next after the blessed bible," Charlotte Elizabeth Tonna wrote, "a flower-garden is to me the most eloquent of books—a volume teeming with instruction, consolation, and reproof"; and "the evening primrose always is, always will be, a momento of what I shall no more enjoy on earth."[24] Botanical moralizing became a part of Victorian popular culture. Readers with an appetite for sentimental flower books could choose among flower poetry, books on the language of flowers, moral and religious works taking flowers as their examples, and books about poetic, folkloric, and historical associations of flowers. Publishers produced many flower books for mass consumption, often combining literary and visual material in a poetic and artistic mélange.[25]

Celebrants of Flora could look to a burgeoning print culture about flowering and nonflowering plants. Writers, journalists, and entrepreneurs fed the popular taste for publications about gardening. Books such as Shirley Hibberd's *Rustic Adornments for Homes of Taste* (1856) gave middle-class readers ideas on how to create their own fernery, heathery, and vinery. Gardening magazines proliferated, in weekly or monthly numbers, cheaply done with wood engravings or prepared with more sensitively rendered color plates.[26] Readers could choose, for example, among the *Botanical Register, British Flower Garden, Floricultural Cabinet, Horticultural Register, Gardeners' Chronicle*, and *Floral World and Garden Guide*, to name but a few.

Within the diverse print culture of early and mid-nineteenth-century botany, periodicals, textbooks, and handbooks multiplied, and writers, editors, and publishers were cognizant of audiences with varied interests and levels of knowledge. Among botanical journals calibrated for different audiences, Benjamin Maund's *The Botanist* (1836–42) set out to combine systematic botanical information and general knowledge, "accurate scientific instruction, with an occasional appeal to the imagination, and to the moral and religious feelings." The editor invoked William Paley's *Natural Theology* (1802) in wanting his articles and illustrations to show "the many testimonies of power and wisdom shown us by the great Creator in the exquisite symmetry and manifest design exhibited in the organic structures of different plants."[27] Similarly, *The Phytologist: A Popular Botanical Miscellany*, situated itself between journals of "a general character" and "those of high scientific pretensions." This was a journal for field botanists that focused on "recording and preserving Facts, Observations and Opinions" about British plants.[28] Writers of books, like editors of botanical periodicals, articulated the audience, direction, size, format, and prices of their botany books accordingly. William J. Hooker produced his *British Flora* (1830) as a single octavo volume, for university students to take out on botanical excursions, because the standard text, Sir James Edward Smith's *English Flora* (1824–28), was too cumbersome to carry and too expensive for students to buy. The "voluminous" size and costliness of Smith's *English Flora* led indeed to the decision to issue

LADIES' BOTANY:

OR

A FAMILIAR INTRODUCTION

To the Study

OF THE

NATURAL SYSTEM OF BOTANY.

BY

JOHN LINDLEY, Ph.D. F.R.S.

ETC. ETC. ETC.

PROFESSOR OF BOTANY IN UNIVERSITY COLLEGE, LONDON.

I boast no song in magic wonders rife,
But yet, oh Nature! is there nought to prize,
Familiar in thy bosom scenes of life?
And dwells in daylight truth's salubrious skies,
No form with which the soul may sympathize?
CAMPBELL.

VOL. II.

SECOND EDITION.

LONDON:

JAMES RIDGWAY AND SONS, PICCADILLY.

John Lindley, *Ladies' Botany, or A Familiar Introduction to the Study of the Natural System of Botany* (1834–37).

a moderately priced second edition in 1832, reducing the text and omitting many illustrative plates. T. S. Ralph's *Elementary Botany* (1849) was written "not to make Botanists of the students or readers . . . , but to convey such an amount of information upon the subject to the generality of learners and especially to the younger portion, as shall be a sufficient guide to them to pursue the study still further." Arthur Henfrey's *Elementary Course of Botany* (1857) targeted medical students.

Publishers increasingly issued books under titles that represented botany as a science; *Elements of Botany for Families and Schools* (1833) was retitled in its third edition *Elements of the Science of Botany* (1837). Popularizations of botany continued to appear, directed to women and children and "adapted to the capacity of the young, consistent with the present state of knowledge, comprehensive in detail, as simple as possible in language, and small in price."[29] Authors, editors, and publishers modified their botany books to fit the temper of the times, as well as to give the representation of the science that they believed in, or believed would sell.

Within early Victorian science culture, new narratives of nature were also being given textual form. Although writers chose diverse pathways for reporting scientific work (e.g., Darwin's travel journals), textbooks and essays began to codify along recognizably "modern" lines, and the standardized scientific text developed during this period as its own genre of expository prose. Writers who demarcated "scientific" practices from "unscientific" or "popular" ones also delineated audiences and styles of knowledge. Characteristic of this direction during the 1830s is David Brewster's preface to a new edition of Leonhard Euler's *Letters to a German Princess on Different Subjects in Physics and Philosophy*. In the 1760s the learned Swiss mathematician and natural philosopher Euler had written letters about natural philosophy to a niece of Frederick the Great; these were published in 1768 and translated into English in 1795 in an edition that carried an extended Enlightenment-style introduction about the importance of substantial education for women. When the book was reissued in 1833, the title was changed to *Letters of Euler on Different Subjects in Natural Philosophy, Addressed to a German Princess*, and the focus of the preface changed too. David Brewster did not discuss the importance of female education but instead focused on how to write elementary and popular science books, juxtaposing old style and new style texts. The old style, he wrote, was "left almost exclusively in the hands of inferior writers, who possess only a general and superficial knowledge of the subjects of which they treat." By contrast, a book in the new style is written by "a philosopher who devotes himself to the task of perspicuous illustration" and whose skills are evident in the "selection of topics, [and] . . . clearness of reasoning."[30] Brewster contrasts, in other words, a scientist's single-minded study of nature and the "superficial" knowledge of popular writers and teachers. The German princess is displaced to the subtitle, and Brewster repositions Euler's book as a new-style scientific text.

Within botany, the textual model of a more masculinized botanical science was shaped during several decades of discussion about how to teach botany and how to write introductory books. Writers began developing textual styles that were less associated with women and home, and books in the familiar format came under attack. John S. Forsyth's *The First Lines of Botany* (1827) illustrates this cultural turn. An exposition of Linnaean ideas with some material on plant structure, it was directed to "the young botanist," who is male. Forsyth demarcated a new narrative terrain in this elementary account. His botany book is not in the familiar format. He omits the mother-educator who, by the 1820s, was a narrative convention in introductory writing. He excludes girls as pupils and also erases any suggestion of a domestic setting. Forsyth does not allow these departures to go unremarked. He has much to say by way of criticizing earlier popular botany books. He specifically excoriates the "familiar style" or familiar format; the familiar style, he writes, is "that which is thrust forward in the conversational or epistolary style, by people indifferently acquainted with the sciences they would thereby promulgate; . . . even in the best hands, there is something in this method at which the unassuming, industrious, and inquisitive mind recoils." Forsyth gave a distinct gender spin to his critique of narrative form in science writing. The "inquisitive mind" recoils at the conversational or epistolary method, he continued, "more particularly, when forced . . . to draw through such equivocal channels, upon the fictitious correspondence of some garrulous old woman or pedantic spinster, for the higher order of elementary knowledge." [31]

Forsyth thus misogynistically stigmatized women science teachers as transgressing his notion of a feminine ideal. A surgeon and medical writer in London during the 1820s and 1830s, Forsyth was given to hyperbolic formulations.[32] His virulence in *The First Lines of Botany* against women popular science writers and women as popular science teachers may stand out as extreme, but it is a voice at the far end of a recognizable spectrum. Forsyth directed his textual misogyny at a narrative practice widespread in women's popular science books. His attempt to reconfigure the style of introductory botany books is one example of backlash in botanical culture during the 1820s and 1830s.

John Lindley also wrote for audiences less well versed in botany and delineated what was appropriately botanical from what, to his mind, did not accord with the "modern science of botany." Insofar as he wanted to rescue botany for science, education was a central part of his broader project. He championed the natural system in his thirty years of teaching at what became University College and in a series of books and periodical publications during the 1830s and 1840s. True to the anti-Linnaean theme of his inaugural lecture, the prospectus for one botany course condemned almost all botanical books published before the nineteenth century and requested students in particular "not to furnish themselves with any introductory work by the late Sir James Edward Smith or Dr. Thornton."[33]

Lindley wrote a series of textbooks meant as alternatives to the Linnaean books and approach that he sought to depose. His *Introduction to Botany* (1832), one of several books designed for students in his classes, explains the natural method of Candolle. He begins with plant anatomy (called "organology"), and proceeds from plant physiology into taxonomy, then into morphology, "by far the most important branch of study after Elementary Anatomy and Vegetable Physiology." Lindley prefaced his explication with a passionate statement about the state of the art of botany. Others had written introductory books about botany, but he winnowed for his readers "the vast mass of evidence upon which the modern science of botany is founded." Two purposes underlay his *Introduction to Botany*: "to draw a distinct line between what is certain and what is doubtful" in botanical knowledge, and to show "a few simple laws of life and structure" in the vegetable kingdom. His *School Botany* (1839) taught the "rudiments of Botanical Science," namely, Candolle's division of plants into exogens, endogens, and acrogans. Lindley directed this explication to young men preparing for university entry, hence the book promises "all that it is necessary to know in order to make a fair beginning."[34] Lindley also produced reference works (notably *The Vegetable Kingdom*, 1846), and wrote most of the articles on botany for the twenty-seven-volume *Penny Cyclopedia* (1833–44), a project developed under the auspices of the radical Society for the Diffusion of Useful Knowledge.[35]

John Lindley's efforts to modernize botany by distinguishing between polite accomplishments and real botany did not mean excluding women as an audience, however. His *Ladies' Botany* (1834–37), a lavishly illustrated two-volume work, is an elementary account of the "modern method in botany" for the "unscientific reader." Based on the epistolary and maternal model of Rousseau, it is organized as fifty letters on structure and the natural system of classification, addressed to a mother who wants to teach her children about plants. *Ladies' Botany*, packed with information about plant structure, is meant to be "an experiment upon the possibility of conveying strictly scientific knowledge in a simple and amusing form." Lindley set out to teach a different language of flowers, in contrast to fashionable and dominant floral discourses of his day. More demanding and intellectually informed, he included no general information about medicinal uses of plants or religious references. His ideal audience con-

BOTANICAL WORKS.

1.

CURTIS'S BOTANICAL MAGAZINE (commenced in 1786); continued by Sir W. J. Hooker, F.R.S. In Monthly Numbers. 6 Plates. 3s. 6d. coloured.

2.

HOOKER'S JOURNAL of BOTANY and KEW GARDENS MISCELLANY. Edited by Sir W. J. Hooker. In Monthly Numbers, with a Plate. Price Two Shillings.

3.

HOOKER'S ICONES PLANTARUM. In Parts, each containing Eight Plates. Price 2s. 6d.

4.

FLORA of NEW ZEALAND. By Dr. J. D. Hooker, F.R.S. Part III. 20 Plates. Price 11s. 6d. coloured; 21s. plain. To be completed in Five Parts.

5.

FLORA OF WESTERN ESKIMAUX-LAND, and the adjacent Islands. By Berthold Seemann. Part I. With 10 Plates. Price 10s. 6d. coloured.

6.

THE VICTORIA REGIA. By Sir W. J. Hooker. With Illustrations of the natural size, by W. Fitch. Elephant folio. 21s. coloured.

7.

The RHODODENDRONS of SIKKIM-HIMALAYA. Thirty coloured Drawings, with descriptions. By Dr. J. D. Hooker, F.R.S. Folio. £3 11s.

8.

A CENTURY of ORCHIDACEOUS PLANTS. By Sir William J. Hooker. Containing 100 coloured Plates. Royal 4to. Five Guineas.

9.

PHYCOLOGIA BRITANNICA; or, History of the British Sea-Weeds; containing Coloured Figures and Descriptions of all the Species. By Professor Harvey. In 3 vols. royal 8vo, cloth, arranged in the order of publication, £7 12s. 6d.; in 4 vols. royal 8vo, cloth, arranged systematically, £7 17s. 6d.

10.

FLORA ANTARCTICA. By Dr. J. D. Hooker. 200 Plates. Royal 4to. £10 15s. coloured; £7 10s. plain.

11.

THE CRYPTOGAMIC BOTANY OF THE ANTARCTIC VOYAGE. By Dr. Joseph D. Hooker. 74 Plates. Royal 4to. £4 4s. coloured; £2 17s. plain.

12.

THE TOURIST'S FLORA. By Joseph Woods. 8vo. 18s.

13.

THE ESCULENT FUNGUSES OF ENGLAND. By the Rev. D. Badham. Coloured Plates. Super-royal 8vo. 21s.

14.

ILLUSTRATIONS OF BRITISH MYCOLOGY. By Mrs. Hussey. Second Series. In Monthly Numbers. Royal 4to. Each containing Three Plates. 5s. coloured.

15.

POPULAR HISTORY OF BRITISH FERNS, comprising all the Species. By Thomas Moore, F.L.S. 20 Coloured Plates. Royal 16mo. 10s. 6d.

16.

POPULAR HISTORY OF BRITISH SEA-WEEDS. By the Rev. David Landsborough. Second Edition. 20 Coloured Plates. Royal 16mo. 10s. 6d.

17.

POPULAR FIELD BOTANY. By Agnes Catlow. Second Edition. With 20 Coloured Plates of Figures. Royal 16mo. 10s. 6d.

18.

VOICES FROM THE WOODLANDS; or, History of Forest Trees, Lichens, and Mosses. By Mary Roberts. 20 Coloured Plates. Royal 16mo. 10s. 6d.

19.

THE CULTURE OF THE VINE. By John Sanders. With Plates. 8vo. 5s.

Reeve and Co., Henrietta Street, Covent Garden.

List of "Botanical Works," *Literary Gazette*, 1853.

sists of "those who from their habits of life and their gentler feelings are the most sensible to the charms of nature" but who want to take their relationship to the vegetable kingdom beyond "a vague sentiment of undefined admiration." So far so good, for giving women readers access to more serious, more "scientific" botany. But this is not a book to prepare women for advanced botanical research. His women are not students themselves but are those who will be teaching their "little people," and there is no suggestion that botanical activity for the ladies will move beyond botany as an "amusement and a relaxation" and cross the threshold into understanding botany "as an important branch of Natural Science." [36] His mother-centered focus reads as less empowering than other female-centered narratives found in the works of many women writers. In fact, although *Ladies' Botany* is Lindley's attempt to modernize botany teaching (and reviewers certainly welcomed it as such), the book belongs to the polite literature of botany, and that is precisely where Lindley would have his ladies remain.

Botany became a discipline in part by means of books and disciplinary treatises that created representations of the field and demarcated one science from another. During the first half of the nineteenth century, entries in encyclopedias such as the *Encyclopaedia Britannica* profiled different scientific disciplines and celebrated heroes who shaped their areas into modern sciences. The seventh edition of the *Britannica* (1842) clearly promoted disciplinary knowledge, dividing the older integrative category of "natural history" into various branches, each with its own boundaries and its own experts.[37]

Why did John Lindley, in 1834, jump into a field dominated by women? Various male writers in the 1830s tried to shape botany as a science rather than an accomplishment, to "redeem" it from its contemporary "obloquy" by guiding it away from "superficial knowledge." This redefinition of botany is not necessarily antiwoman, although Lindley's contrast between botany as "an amusement for ladies" and as "an occupation for the serious thoughts of man" clearly narrows the space for female botanists. In *Ladies' Botany* he joined the cadre of botanical writers who addressed a female audience while also constructing an ideal reader for popular botany books. Forsyth's introductory book from 1827 castigated popular female writers as "pedantic spinsters," and Lindley, committed to a view of the field as serious, seems to want to reclaim male authorship of botany books for women. In a sociological study of literary production during the Victorian period, Gaye Tuchman documented how Victorian literary culture "edged out" women novelists by raising the status of the novel and elevating the novelist into a white-collar occupation.[38] The profession-alization of botanical science and the elevation of the nonamateur male botanist are counterparts to this direction.

During 1830–60, botany was increasingly shaped as a science for men, and the "botanist" became a standardized male individual. Botanical dis-course devalued the familiar format and set its course toward a deperson-alized and standardized text. The standard science textbook is accordingly a gendered text and by no means a neutral textbook style. Botanical science moved away from a domain of some discursive female authority.

And what became of the familiar format? The genres of conversation and letters retained their publishing power in expository science well into the nineteenth century in books for children, women, the uniniti-ated, and the "amateur." Family rhetoric continued to be a literary means for offering correction and guidance. Letters and conversations remained narrative forms for drawing readers over the threshold and for teaching elementary topics. Justus von Liebig, for example, chose the epistolary format for his *Familiar Letters on Chemistry* (1843–44), written for public education. The conversation was also taken up as a deliberately nostalgic form by Victorian writers resistant to contemporary directions in sci-ence, as in Charles Kingsley's *Madam How and Lady Why* (1869). But the familiar format came to be represented and seen, among other things, as "feminine," not serious, nonscientific, and also old-fashioned in its con-nection of science and religion. Dialogues, Greg Myers wrote, "became increasingly marginal to the business of scientific education." [39] Academic botanists and professionals did not write scientific dialogues.

Formal changes in textual styles for botany books came from a series of changes within botanical culture, as indeed in science culture more gen-erally. The science textbook became codified along different lines. Instead of the familiar format, the catechism became the model of narrative and pedagogical practice in science at the turn of the nineteenth century. A narrative form transposed from theology into didactic writing, the cate-chism—a model of set questions and answers—became the hallmark of a system of education associated with boys' schooldays and rote learning. This method, suited to the Information Age of the late Enlightenment, was widely used for teaching many subject areas. William Mavor and William Pinnock were particularly well known as authors and publishers of catechisms on science topics, notably Mavor's *Catechism of Botany* (1800) and Pinnock's *Catechism of Electricity* and *Catechism of Mineralogy* (ca. 1830). The catechetical method teaches through rote recitation and appeal to authority and offers quick access to information. It was seen as a tool for the harried teacher. Richmal Mangnall's *Historical and Miscellaneous Questions for the Use of Young People* (1800), for example, was written for pupils "in public Seminaries" whose teachers did not have "sufficient leisure to converse with each separately."[40] The catechism, by teaching science without personalities and without context, embodies a learning pattern different from the familiar format. It represents instruction rather than education and school culture rather than home and leisure, a world apart from participatory teachers and from mother.[41] The catechism, a third-person exposition with no narrative dimensions, set the stage for teaching science without personalities and without context. This is the direction of the standardized text format. Over the course of the nineteenth century, the spread of formal schooling took science study away from home. Professionalizing modes in science culture led to hierarchies about different kinds of learning and different kinds of texts. Debates about pure and applied science affected links between science education and general education.

Within the history of textbook writing, generic differences between introductory botany books from the 1790s and the 1860s illustrate broader cultural shifts from an integrative general education approach toward specialization. Priscilla Wakefield's *Introduction to Botany* (1796) and Daniel Oliver's *Lessons in Elementary Botany* (1864) will serve as examples. Wakefield's book was a basic introduction to botany for two generations of readers. It was, as discussed in chapter 4, an epistolary account that situates botany as an improving home-based activity shared between sisters and as part of valued family routines. She took science education seriously for young women, and her botany book embodies an Enlightenment approach to the moral and intellectual value of natural knowledge. It is a family-centered book of education that teaches more than botanical systematics. Oliver's book, which was the standard introductory botany text for a later generation, illustrates the genre of science writing that became conventional during the later nineteenth century and thereafter. *Lessons in Elementary Botany* was meant to be introductory to excellent textbooks of botany already published, "to insure to the diligent learner a sound foundation for more advanced Botanical studies."[42] In expository chapters about structural and physiological botany, the author discusses classification, using natural orders. The book is addressed to a non-gender-specific "you," with a few references to the "learner" as "he." There are no personalities, and no general lessons about life beyond the specifics of botanical instruction, or interjections with miscellaneous information. Oliver offers no exhortations about the value of botanical study for one's intellectual, moral, or spiritual well-being. *Lessons in Elementary Botany* is a book of instruction rather than a book of education. It is formal and nonfamilial, depersonalized and decontextualized, presenting science study as separate from moral education and from everyday life.

Oliver's book is a revision of a manuscript that J. S. Henslow, Charles Darwin's tutor and mentor and professor of botany at Cambridge, left unfinished at his death. There Henslow set out ideas about botany teaching that he had put into practice when, during the 1850s, he conducted a school for the sons and daughters of laborers in the Cambridgeshire village of Hitcham. He reported on his methods and goals in a series of essays in the *Gardeners' Chronicle* of 1856, characterizing botany "as a branch of a general liberal education," and promoted the power of a classificatory science as tending "to strengthen the reasoning powers." Henslow distinguished "conveying useful instruction" from educating "the mental faculties, and improving the moral apprehensions of young children." He was interested in "the way botanical knowledge relates to everyday life" not as training for specialization but instead as mental training.[43]

It is striking to turn from these ideas to the revision of Henslow's manuscript by Oliver (professor of botany, University College London). *Lessons in Elementary Botany* contains no trace of the generalist emphases of Henslow's own teachings. Oliver omitted Henslow's material about botany as an "educational weapon" as well as such informational material as Henslow's discussion of beermaking in relation to hops. He addressed these omissions in his preface and explained them in terms of both cost of the volume and curricular fluidity; teachers, he suggested, can supply the applied material themselves. A broader cultural reading seems warranted, however.

By the 1860s, botany in England was being given a more modern profile. The observational and field tradition of natural history and the research and laboratory tradition of botany were going their separate ways. The site of legitimation for botany was soon to become the laboratory, and the old natural history, based in systematics and the field tradition, was superseded by the new biology. As the place of science was defined, "militant anti-amateurs" became the ideal professional practitioners.[44] Geographical and conceptual shifts in botany after the 1860s increasingly excluded amateurs, working-class botanists, and women from the mainstream of high botanical science. Science education was a topic of national concern in England at that time. An inquiry into endowed and grammar schools included discussion of how to promote science in schools; the 1867 Report of the Taunton Commission led to shaping botany as a school subject, one that could train pupils in the methods of science. There was, as a result, considerable discussion about how to teach botany and write textbooks to represent new approaches to science education. Teachers and writers sought to separate their model of school science from the home-based general education science practice that women's popular science narratives represented.

From 1760 to 1830, the gendered shape of botanical culture gave women access to botany, but after 1830 the same gendering was inverted to deny them access. Women elbowed in but then were elbowed out. The gendered form of botanical culture shaped the social practices of most women in botany. It shaped what women learned, how they practiced science, and where they could pursue science. Over the ensuing years, while botanical brothers (like the brother of Sarah Fitton) completed medical studies and then established a place in the public world, perhaps with a gentlemanly avocation for natural history, sisters were in all likelihood at home, perhaps studying plants with their microscopes, perhaps corresponding with clergymen-naturalists, perhaps engaging with botany as part of polite culture, perhaps moving more deeply into plant physiology, or perhaps, having married and become mothers themselves, teaching botany to their children. Their botany was in the breakfast room, in a separate sphere. Nevertheless, within the vocabularies of early and mid-nineteenth-century gender ideology, women took part in botanical activities of many kinds and continued to make botany a resource for themselves as research assistants, artists, teachers, collectors, correspondents, and also writers.

Women and Botany in the Victorian Breakfast Room

ROSEBUD OF ENGLAND! young and cherished Flower,
    To thee a sisterhood of Flowers we bring,
    And pray thee to receive the offering
Which simple Song, and Art's most gentle power,
Have blent to form for thee.

> Dedication to Her Royal Highness the Princess Victoria,
> Louisa Ann Twamley, *Flora's Gems*, 1837

The sea-weed collector who has to pick her way to save her boots will never be a loving disciple as long as she lives! Any one, therefore, really intending to *work* in the matter, must lay aside for a time all thought of conventional appearances. . . . Next to boots comes the question of petticoats; . . . to make the best of a bad matter, let woollen be in the ascendent as much as possible; and let the petticoats never come below the ankle.

> Margaret Gatty, *British Sea-Weeds*, 1862

E ven as botanical science was being shaped as something distinctly different from polite science, botany continued to be promoted as an endeavor that taught moral and religious lessons, prevented idleness, and was a rational recreation for both men and women. The evangelical revival of the early nineteenth century linked religious beliefs and practice with domesticity and family, and also with doctrines about both manliness and femininity. The new serious Christianity mandated particular kinds of pursuits for middle-class women, and botany was not at odds with teachings about women's gendered roles and appropriate activities.[1] Indeed, the study of nature was congruent with significantly heightened emphases on moral regeneration through religious beliefs and practices. Discourses of femininity that identified women with flowers helped smooth their path into botanical work. As one writer remarked in 1838: "That the mental constitution of the fair sex is such as to render them peculiarly susceptible of whatever is delicate, lovely, and beautiful in nature and art cannot, we think, be controverted; we are not, therefore, surprised that Botany receives more of their attention and study than any other science."[2] In line with beliefs like these about women's nature and separate spheres for women, botany had a socially sanctioned place in the early Victorian daily round, at home in England as also in the colonies.

During the early nineteenth century, popular botany was largely a safe and unproblematic part of education and leisure for women in the middle classes. Women collected plants, developed herbaria, continued to study Linnaean botany, examined botanical specimens under microscopes, and drew and painted flowers. Their botanical activities served a variety of personal and family interests. Botanical work also fulfilled broader educational, social, and religious responsibilities assigned to women within the gender economy of those decades. Women in nineteenth-century England were widely taught that benevolence was part of their nature and their mission, and thus they were exhorted, as teachers and as mothers, to take up the vocation of philanthropy. Thus women were busy as charitable visitors, sick visitors, and prison visitors, organizing bazaars, missionary societies, and other voluntary societies, and subscribing to philanthropic causes. The antislavery movement was another prominent route for female philanthropic activism, and women in groups such as the Ladies Anti-Slavery Association gathered signatures on petitions, encour-

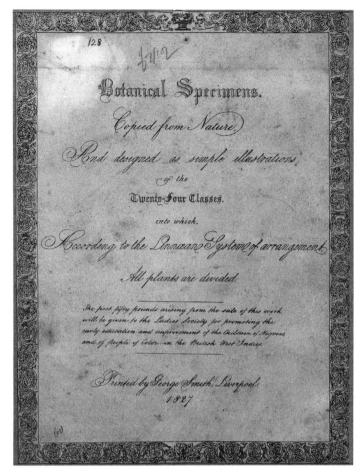

Botanical Specimens Copied from Nature (1827). (Courtesy of Hunt Institute for Botanical Documentation, Carnegie Mellon University, Pittsburgh)

aged boycotts of Caribbean products, and collected subscriptions toward the abolitionist campaign during the years before the emancipation bill passed.[3] One women's antislavery association in the late 1820s made botanical culture work for them by using botanical art as a fund-raiser; part of the proceeds from the sale of *Botanical Specimens Copied from Nature* (1827) were intended for "the Ladies' Society for promoting the early education and improvement of the Children of Negroes and of People of Color in the British West Indies."

For the most part botany for girls and women was based at home in the parlor or the breakfast room, or near home. The field tradition of botany also guided women out on local excursions in search of flowering plants or fungi, and even onto the rocks and into the pools of the seashore. Additionally, women attended public lectures and semipublic *conversazioni* that were open to them. Some girls and women of the laboring classes had access to botanical study too, as part of artisanal botany in pubs, or in rural villages learning under enlightened pedagogues.[4]

As the social geography of early Victorian botany became more sharply demarcated, however, women were increasingly kept on the outside of mainstream institutions. Earlier in the century the Royal Institution in Albemarle Street, London, a "stronghold of fashionable science," had cultivated female subscribers and organized its activities with an eye to its many female members.[5] By contrast, the foundation years of the British Association for the Advancement of Science (BAAS) illustrate changing directions in science culture. During the 1830s, when the newly established association set out to enhance the status of British science and to interpret science for intelligent laypeople, women attended meetings and by their participation contributed to its financial viability. Only men, however, were asked to sign the founding document of the BAAS, and battle lines soon were drawn between those who wanted to ensure a place for women within the organization and those who argued that a visible female presence would prejudice its serious profile. Thus the conservative (and creationist) president-elect William Buckland opposed women's involvement when the BAAS met in Oxford in 1831 and explained his position this way: "Everybody whom I spoke to on the subject agreed that if the Meeting is to be of Scientific utility, ladies ought not to attend the reading of the papers — especially in a place like Oxford — as it would overturn the thing into a sort of Albemarle [Royal Institution] dilettante meeting instead of a serious philosophical union of working men." Thereafter women were an important part of the public rationale for the organization, "central to [its] style and success," but were "irrelevant to its manifest purposes and debarred from any formal say in its government."[6] Among other formal institutions of science and botany, women had no access to the Linnean Society and the Royal Society, both of which remained closed to them until the twentieth century.

The egalitarian Botanical Society of London was the one striking exception to the professionalizing direction within botanical institutions at that time. This progressive and distinctly metropolitan society counted liberals and intellectuals of many kinds among its members. It had a deliberately nondiscriminatory membership policy, and women made up

10 percent of the members. Yet that worthy institution was unable to recruit more women.[7] Even there, where women could cross the threshold of institutional membership, they did not advance from that one vestibule into the heart of the enterprise of mid-nineteenth-century botanical science.

This is not surprising, for social codes during the early Victorian decades circumscribed women's sphere, the sexual sciences proffered explanations of women's nature, and prescriptive literature of many kinds laid out gendered parameters of women's lives that were more restrictive than during the Enlightenment. Women botanized for the most part within dominant Victorian ideas about sexual difference, separate spheres, and gendered activities. The editor and botanical-horticultural writer Jane Loudon, for example, opened the first issue of her midcentury magazine the *Ladies' Companion* with a statement about women's "nature" and "duties." She wrote: "The necessity of mental cultivation will be strongly enforced; not to make women usurp the place of men, but to render them rational and intelligent beings. The paths of men and women are quite different; and though both have duties to perform, of perhaps equal consequence to the happiness of the community, these duties are quite distinct."[8]

Nevertheless, although a domestic middle-class ideology may have reigned, the Victorian gender economy was not uniformly rigid. There was polyphony of practices within teachings about women's roles.[9] Botany remained a multifaceted opportunity for girls, young women, mothers, and older women during the early Victorian years. Home based, part of education and family activities, it was an intellectual resource for the noted mathematician and woman of science Mary Somerville, for example, who, as a young mother nursing a baby, "devoted a morning hour to the science." "I knew the vulgar name of most of the plants," she reported, "now I was taught systematically, and afterwards made a herbarium, both of land plants and fuci."[10] Emily Shore grew up in a family that valued incremental knowledge and supported a daughter's studious interests. The young and gifted Emily Shore (1819–39) was already a botanophile by age twelve, able to describe and classify plants according to the Linnaean categories. Journals record her work in systematic botany, starting with taxonomy (botanical names and descriptive details) and then shifting into study of the habits, properties, and uses of plants. Even when poor health curtailed her undertakings, Shore continued with some botanical work, and by 1836 she had amassed hundreds of ferns, which she sewed and pasted into portfolios. She wrote in her journal on October 1, 1836: "I finished this evening all that I intend to do in the arrangement of the Penang ferns. It has been an intense labour. I have worked at it from morning till night, with scarce a moment's intermission. . . . The ferns amount to many hundred in number, and are of all sizes, from the length of four feet five inches to two inches. . . . Each fern I fasten in by sewing and gumming." She had a sense of herself as a student apprentice to the science of botany, trying on the botanist's authoritative voice. Thus we hear the aspiring botanist when she discusses plants seen on a walk:

> There are two sorts of *Hippuris* (mare's-tail). Withering and Galpine only mentioned one; the other is, according to Rousseau, not much known. The first I am well acquainted with. It is very much smaller than this, which may possibly be only a large variety of it; but, then, I have never seen the two sorts growing together, and they are certainly very different, so I hope this is the rare kind. But, unfortunately, I know of no book which describes it, for Rousseau only just mentions its existence.[11]

This is not the kind of botany that Charles Lamb and some other male romantics approved of, but it is a way for an intellectual adolescent girl to claim space for herself through language. She is exactly what every mother in the familiar format would wish her daughter to be by the end of their botanical conversations.

As in previous Linnaean generations, Victorian family networks continued to bring girls and women into botanical pursuits, particularly into networks that helped to cultivate their interests. Thereza Mary Dillwyn Llewelyn, in the tradition of Linnaeus's own daughter Elizabeth Christina Linnea, compiled a herbarium and wrote a botanical report that was read at the Linnean Society in 1857, when she was twenty-three years old. She was the daughter and granddaughter of botanists who were fellows of the Linnean Society and was encouraged by the botanist George Bentham.[12] In a similar vein, Anna Children Atkins (1799–1871) grew up within a scientific household as the daughter of the scientist J. G. Children, who had a long career in the British Museum, latterly as the first keeper of the Zoological Department. She was the only child of a devoted and supportive father widowed at her birth. A skilled artist, in 1823 she illustrated her father's translation of Lamarck's *Genera of Shells*, and various of her drawings were published in that decade. Through her father she came into contact with William Hooker, who became her botanical mentor and tutor. J. G. Children was vice president of the Botanical Society of London for some years, and Anna Atkins was also a member of this important egalitarian institution for field botanists. The social and scientific lives of daughter and father remained closely entwined. He was her channel for scientific communication, and in later years he lived with her and they took up photography together.[13]

The tradition of the female helpmate, idealized and sentimentalized in various cultural discourses and influentially embodied in Victorian England as the angel of the hearth, was active in botanical culture too. The working relationship between Anne Elizabeth Baker (1786–1861) and her older brother, for example, illustrated an idealized brother-sister companionship that was a theme in juvenile fiction and verse. Anne Elizabeth Baker took on many research tasks for George Baker's *History and Antiquities of the County of Northampton* (1822–41) and developed expertise in philology, geology, and botany, including compiling plant lists for many parishes, preparing botanical notes, making drawings, and engraving plates. She was celebrated in a contemporary journal not for her own achievements, but for her dedication to his project, as "companion of her brother's journeys, his amanuensis, his fellow-labourer . . . [riding] through the county, by her brother's side."[14]

Artists

During this period women were most visible in botanical culture for artistic work. Middle-class girls, excluded from most occupations by social norms and also from most formal institutions of science, found ready encouragement to draw and paint flowers. Botanical drawing and flower painting enabled women to cultivate aesthetic skill and scientific study within the confines of home and family. The taste for artistic representation of flowers was nourished by magazines, books, and lessons for female audiences of various ages. Miscellanies such as *Elegant Arts for Ladies* (ca. 1856) gave instruction about applied floral arts, among them featherwork, beadwork, waxwork, and dried-flower work. *The Floral Knitting Book* (1847) taught how to knit "Imitations of Natural Flowers," and the author, using botanical language for parts of flowers, described techniques for knitting fuchsia, narcissus, a snowdrop, and nine other flowers using wool and wire. The "Inventress" advertised lessons ("Ladies can take either a Single Lesson, or a Course") in this "elegant and interesting amusement," meant as an activity for the drawing room.

Flower painting, a conventional skill for women of leisure during the eighteenth century, continued as part of the polite culture of sociability. Two generations of women in a family of landed gentry in Gloucestershire completed three hundred paintings of local wildflowers during the 1830s and 1840s. Their "Frampton Flora," a family flora, was a collective project done by at least eight women in the Clifford family, who took botanical excursions near their home villages in the Severn Vale and made field sketches of local flora. They completed watercolors at home and appended basic Linnaean information. The result, an informal local flora, memorializes those many unpublished albums of flower paintings in Victorian England completed by women who joined art and botanical knowledge in an activity fully concordant with gender ideology for women of leisure.[15]

In botany, as in other areas of women's work during the nineteenth century, women were part of "the Hidden Investment," the economic enterprise that harnessed family patterns, business practices, and religious beliefs and contributed invisible and unpaid labor of many kinds to family businesses.[16] As an extension of their responsibilities as wives, mothers, daughters, and sisters, women contributed their artistic work to botanical family firms. Like Ann Lee during the 1760s and 1770s completing botanical drawings for her father James Lee at the Vineyard Nursery and for his friends, women in the nineteenth century were a part of Victorian botanical cottage industries. Women within the Sowerby family continued the tradition of botanical illustration established by James Sowerby, who was responsible for the drawings in the thirty-six-volume *English Botany* (1790–1814); Ellen Sowerby (ca. 1810–63) and Charlotte Caroline Sowerby (1820–65) both were botanical artists. Miss S. Maund contributed many drawings, paintings, and engravings to her father's periodical *Botanic Garden*, and to Maund's and J. S. Henslow's *Botanist* during the late 1830s. John Lindley's two daughters also prepared illustrations for his books. Women in the extended families of J. S. Henslow, Dawson Turner, William J. Hooker, and his son Joseph D. Hooker—four men at the heart of nineteenth-century botany—also assisted over several generations. Mary Turner (1774–1850), a skilled and highly trained artist who studied drawing and etching for many years, prepared the plates for her husband Dawson Turner's monograph on British seaweeds, *Fuci* (1808–19). Her daughters Maria and Elizabeth took art lessons and drew and engraved for William Hooker. Maria Turner (1797–1872) married William Hooker and served for fifty years as her husband's amanuensis and helper, first at the university in Glasgow and later at Kew. Their son Joseph's first wife, Frances Henslow (1825–74), was a daughter of J. S. Henslow, professor of botany at Cambridge, and she followed a family practice by assisting her husband with his botanical correspondence and publications.[17] Another Henslow daughter, Anne Henslow Barnard (d. 1899), contributed drawings to the *Botanical Magazine*, and her drawings were used for the woodcuts in Daniel Oliver's *Lessons in Elementary Botany* (1864), a well-known introduction to systematic botany based on Henslow's lectures. Family projects offered these women a means of developing their own skills.

Emily Stackhouse (1812–70), a skilled botanical watercolorist who was responsible for many illustrations in popular and widely reprinted mid-nineteenth-century books about native plants, straddled the culture of accomplishments and the culture of art as a profession. From a family of botanists and scientists, she collected mosses, grasses, and flowers in Devon and Cornwall and was a member of the Royal Institution of Cornwall in Truro and the Royal Cornwall Polytechnic Society in Falmouth, where she entered collections of plants into exhibitions. From 1830 to 1835 she produced 600 delicate watercolors of plants. Emily Stackhouse was one of several women whose skills contributed to botany books compiled by Rev. C. A. Johns, including *Forest Trees of Britain* (1847–49), *A Week at the Lizard* (1848), and *Flowers of the Field* (1851). She prepared 250 woodcuts for his books, and these were reprinted in at least five other publications issued by the Society for Promoting Christian Knowledge. The amplitude of her artistic work suggests a serious commitment to botanical art, yet her contributions were never formally acknowledged. She probably worked as an amateur within provincial culture, neither seeking nor receiving economic reward for her botanical art.[18]

For women based in the metropolis, floral art could serve as a deliberate means to income. Augusta Innes Withers (ca. 1793–1860s), well-known professional illustrator for floral magazines and "flower painter in ordinary" to Queen Adelaide, also gave lessons in flower painting. Emma Peachey, wax-flower maker to Queen Victoria, produced enormous bouquets for public display and invited commissions from wealthy patrons. She produced ten thousand wax white roses for Queen Victoria's wedding. Emma Peachey held daily viewings at her London residence to promote her art. Her *Royal Guide to Wax Flower Modelling* (1851) gave instruction in "an accomplishment at once royal and feminine in its origin and progress." Emma Peachey was an army officer's daughter compelled by "a change of circumstances . . . to make the art of wax

Frontispiece to Emma Peachey's *Royal Guide to Wax Flower Modelling* (1851). (Courtesy of Hunt Institute for Botanical Documentation, Carnegie Mellon University, Pittsburgh)

flower modelling a source of profit." Alert to commercial opportunity, she promoted wax flowers for use in art classes and as illustrations for botanical lectures.[19] For Mary Harrison (1788–1875), flower painting was a family business in a matrilineal pattern, for she trained her daughters Maria and Harriet to continue her work. Mary Harrison drew orchids for the *Botanical Magazine* and helped form the New Society of Painters in Water-Colours.[20] The tradition of fruit and flower painting may have been a "lowly genre" within academic discourses on art, but women such as Annie Feray Mutrie and her sister Martha Darley Mutrie had successful practices in London during the 1850s; as artists they were "commended for their botanical accuracy as much as for their aesthetic properties." [21]

Collectors and Correspondents

National and colonial botany alike drew on the enthusiasms of women who botanized, often with lists of requested specimens in hand. During decades when imperial travelers sent Wardian cases of exotic specimens to the Royal Botanic Gardens, naturalists at home in England collected, named, described, and glorified indigenous flora. The romance of nature emphasized native English plants, and many female enthusiasts joined in a fashionable hunt for botanical specimens; one botanical neophyte described herself in 1836 as "moss mad." [22] Others took active part in the "fern craze" of the 1840s. The "first woman pteridologist," Margaretta Riley (1804–99) construed botany as a science rather than a fashion and studied native and hardy species of ferns. She and her husband collected, cultivated, and classified ferns from the 1820s until the mid-1840s. Margaretta Riley joined the Botanical Society of London in 1838 (her husband had joined one year before), wrote two papers for presentation there, prepared a jointly written monograph on growing British ferns, and developed a collection of every species and variety of fern in the British flora.[23] Many women contributed to nineteenth-century floras by supplying information about specimens and localities to those cataloging projects. Compilers customarily acknowledged those who had formed collections of local plants; Anne Frances Wingfield (1823–1914), for example, collected plants and recorded observations during the 1840s and 1850s, and her name "appears as authority for the first record of a very large proportion of the Rutland Flora." [24] The *Phytologist*, a mid-nineteenth-century journal for field botanists associated with the Botanical Society of London, regularly published letters from women reporting on their findings of plant species and on plant localities. The opening volume in 1841, for example, recorded Anna Worsley's communication "New locality for Carex clandestina" and Anna Carpenter's "Note on Irregularities in the Flowers of Tropaeolum atro-sanguineum."

The field tradition was dependent upon good relations between collectors in the field and individuals who needed to receive specific plants from local areas and willingly handed out assignments to enthusiastic botanizers. Sir William Hooker, as professor of botany in Glasgow, compiler of *British Flora*, and later director of the Royal Botanic Garden at

Kew, corresponded with many collectors about sending him specimens. Women were part of his circle. Correspondence networks like these were important sources of botanical information and actual specimens. Thus Margaret Stovin (1756–1846), who collected plants in South Yorkshire and Derbyshire for over four decades, had a wide circle of correspondents and exchanged specimens and information with botanists such as Sir James Edward Smith, John Stevens Henslow, and Sir William Hooker. She developed a British flora that contained nineteen hundred specimens and asked Hooker's advice about where to deposit her herbarium.[25]

Anna Worsley Russell (1807–76), another plant collector, was "the most accomplished woman field botanist of the day." From a prosperous and intellectual Unitarian family in Bristol, she contributed records to several local floras. She drew up a list of flowering plants of the Bristol area for H. C. Watson's *New Botanist's Guide* (1835), for example, and later compiled "Russell's Newbury Catalogue," a list of plants found in that area, for *The History and Antiquities of Newbury and Its Environs* (1839). She corresponded with leading botanists such as the notoriously hard-to-satisfy Watson and was a member of the Botanical Society of London from 1839–41 onward, participating in its system for distributing duplicates of herbarium specimens to other members. She sent notes to the *Phytologist* and the *Journal of Botany* about her plant finds and pursued a special interest in fungi, of which she completed over seven hundred drawings.[26]

During this period, several women were well known as seaweed collectors. In a cultural genealogy of female botanists, lady algologists are a notable subset. Most contributed to knowledge of plant groupings by collecting marine plants and sending them to botanists who were compiling large synoptic works on seaweeds. The celebrated Amelia Griffiths (1768–1858), for example, worked in Torquay, Devon, and sent specimens for Dawson Turner's *Fuci* (1808–19) and William Harvey's *Phycologia Britannica: A History of British Sea-Weeds* (1846–51). Dawson Turner named the *Fucus griffithsia* in her honor, and William Harvey wrote that *Mesogloia griffithsiana* "worthily bears the name of its discoverer, . . . who has added so many original observations on the British Algae to the common stock, and has been the first to notice so many new species."[27] Charles Kingsley lionized Griffiths in *Glaucus, or The Wonders of the Shore* as part of his construction of the naturalist as a heroic explorer. He referred to her vigor and dedication and added that English marine botany "almost owes its existence" to her "masculine powers of research."[28] She was as-

Margaret Gatty (1809–73). (Courtesy of Hunt Institute for Botanical Documentation, Carnegie Mellon University, Pittsburgh)

sisted, and then succeeded, by her daughter Amelia Griffiths (1802–61). Their botanical household also included Mary Wyatt, a former servant of Mrs. Griffiths, who compiled (and sold) volumes of Devon marine algae; Mary Wyatt sold *exsiccati*, mounted specimens of actual plants. Her *Algae Danmonienses* (1835) consisted of two hundred examples of marine plants collected in Devon.

Margaret Scott Gatty (1809–73), another well-known member of the seaweed sorority, was introduced to the study of botany as a health regimen when, middle-aged and suffering from the effects of too many pregnancies, she moved from Liverpool to the seaside in 1848. She brought a no-nonsense robustness of spirit and practice to her work on marine plants. For years she led family expeditions in search of seaweeds; her daughter, later the writer Juliana Ewing, shaped a verse parody of her mother under the title "At Home and at Sea—A Ballad":

O Gattys! go and call your mother home
    Call your mother home
    At least in time for tea!
The breakfast, lunch, and dinner go and come
    Unheeded, at the sea.

The creeping tide came up along the sand,
    And round and round the sand,
    But not a step moved she.
Her children shouted to her from the land
    She shouted to the sea.

"O! is it weed or fish or floating hair?
    A zoophyte so rare,
    Or but a lump of hair,
    My raptured eyeballs see?
Were ever pools so deep or day so fair—
    There's nothing like the sea!

"My ungloved hands grasp Laminaria's root,
    A crab crawls o'er my foot,
    I would I were a brute.
    I hate society.
The water gurgles sweetly in my boot!
    There's nothing like the sea!"

They saved her life by force and dragged her home.
    She vowed in time to come.
    So far she would not roam;
    But vainly promised she.
For still at night the Gattys call their mother home,
    And save her from the sea.[29]

Margaret Gatty, a clergyman's wife, was in many respects a conventional Victorian woman of her class and station. But she was also a Victorian botanical bacchante who trekked along the shore in an unconventional collecting costume she devised for herself. She accepted that women of her time and class were inevitably wrapped in "draperies" as the costume for botanical work. But she stipulated the fabric and length of petticoats (see epigraph to this chapter), advised against cloaks and shawls, and called for "hats" rather than "bonnets," and "a strong pair of gloves": "all millinery work, silks, satins, lace, bracelets and other jewellery etc. must, and will be, laid aside by every rational being who attempts to shore-hunt."[30] Mrs. Gatty's "shore-hunting" was her hobby, an avocation that ran parallel to her other life as "Aunt Judy," writer of children's books. Yet her absorption in her chosen algological projects, her long-term correspondence with male experts, and her authorship of the two-volume *British Sea-Weeds* (1862) showed a focus that clearly went beyond accomplishment, sociability, and ordinary leisure.

Elizabeth Andrew Warren (1786–1864) lived on the coast of Cornwall within view of Falmouth, and her main occupation was said to have been "making and arranging her various collections." She amassed lichens, mosses, ferns, and heather but was particularly dedicated to "the growth, not of the land, but of the sea" and was one of the many women who sent specimens to metropolitan botanists compiling large-scale floras. Her interest in collecting plants began in her childhood; regarding a plant that was "a primeval favourite of mine," she recalled "one of my first summers [when] I used to *crawl* on the banks to pluck it."[31] During her adult years, family naval connections provided her with a steady supply of exotic specimens from the Crimea, the East Indies, Hong Kong, and the Falklands. Her special interest, however, was Cornish plants. She compiled a three-volume collection of native plants, named and arranged according to Linnaean categories, and presented the "Hortus Siccus of the Indigenous Plants of Cornwall" to the Royal Horticultural Society of Cornwall. She discovered several new species of seaweed (one of which was named *Schizosiphon warreniae* in her honor) and wrote about her finds.[32]

Elizabeth Warren's interest in collecting plants led her in 1834 into a correspondence with William Hooker that lasted nearly twenty-five years. She first wrote to him in Glasgow, where he was the Regius Professor of Botany, to express her admiration of his *British Flora* (which, she explained, she used in arranging the plants in the "Hortus Siccus") and to ask how to locate a book mentioned in his introduction. Over the years they exchanged live plants and seeds and discussed identification of specimens. She offered to send him plants from Cornwall, listing extensive

Herbarium specimen of a "plantago" collected in June 1841, in Elizabeth Andrew Warren's "Hortus Siccus of the Indigenous Plants of Cornwall." (Royal Institution of Cornwall)

possibilities for his selection, and reported asking "the Ladies of Penzance for some of the rarities of that neighbourhood" for him.[33]

In addition to contributing to botanical culture by her collecting work, Elizabeth Warren stepped briefly, and painfully, into the text culture of nineteenth-century botany when she was about fifty years old. In the 1830s she prepared and published *A Botanical Chart for Schools*, a wall chart about the Linnaean system. Teaching by charts was popular at that time, but a specifically botanical chart had not yet been tried. Her pioneering venture was a colored chart on canvas, varnished and hung on a roller, intended for an audience of younger pupils. The project had a family genesis. "It happened," she explained to Hooker, "that the Governess who has had the care of my Neices [*sic*], expressed a strong wish to have the first rudiments of Botany, as she had other parts of Natural History, in the manner of Charts for the instruction of her pupils; and which she had found very useful in her School."[34] Warren arranged botanical information in columns that displayed characters of each Linnaean class, defined the orders, and enumerated the genera; she also included observations on the uses of plants, "to relieve . . . the dry detail of long words." Recollecting herself as a botanical novice, Elizabeth Warren shaped her chart for pupils at the elementary level of knowledge and limited the amount of information to be included. She took as her model two books written for the level of audience she envisaged for her work: Sarah Fitton's *Conversations on Botany* (1817) and Caroline A. Halsted's *The Little Botanist* (1835).

William Hooker had not seen a botanical chart before and enthusiastically supported her project. Over the course of their correspondence, he assisted her substantively and editorially and commiserated with her about problems with printers and booksellers, distribution and sales. She had prepared two different charts and paid fifty pounds herself for one or both to be printed, but few copies sold; she admitted that she published "in total ignorance of the mighty power and influence of the Booksellers."[35] Hooker tried to help by arranging for notice of the work to appear in Taylor's *Annals and Magazine of Natural History*. Hooker himself may have written the review in 1839 that referred to Miss Elizabeth Andrew Warren's "Botanical Chart for Schools" as an "excellent Chart . . . [which] ought to be in the hands of every teacher of youth throughout the kingdom" and praised Warren as "a lady whose accurate researches in British botany have obtained for her a name which will rank with those of Miss Hutchins and Mrs. Griffiths (and we scarcely know if, botanically speaking, we can pay her a higher compliment)."[36] Over the next

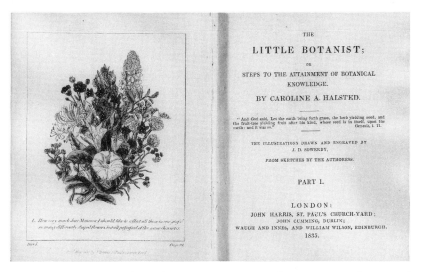

Caroline A. Halsted, *The Little Botanist, or Steps to the Attainment of Botanical Knowledge* (1835). (Osborne Collection of Early Children's Books, Toronto Public Library)

several years, Elizabeth Warren grappled with the possible reasons her chart did not sell better. Chronicling for Hooker "the tormenting subject of my Chart publication," she speculated that the title *Botanical Chart for Schools* was a mistake and deterred librarians from purchasing it. She discerned little interest on the part of schools in receiving botany "as a branch of general education." She suggested that masters and mistresses were reluctant to teach what they had not studied themselves. Because she thought her chart would be serviceable for lecturers and students in medicine, she tried unsuccessfully to arrange for it to be reviewed in the *Lancet*. Finally, out of her own commitment to the broader learning possibilities in botany, she donated fifty remaining copies of her chart to schools in 1844.

Elizabeth Warren created a pioneering visual aid in her botanical chart and contributed small pieces to a local publication, but she ventured no further into scientific botany. Although well versed in local botany, she never compiled her knowledge into a larger work. For reasons that are unclear, she did not join the Botanical Society of London, a nationwide group that sought out women members during the late 1830s. Had she been a man, her collection of Cornish plants might have given her access to membership in the Linnean Society. During this same period Rev. Charles A. Johns, a schoolmaster in Cornwall, also corresponded with Hooker and sent him plants. He entered exhibitions of the Royal Cornwall Polytechnic Society in Falmouth, won some awards, and recommended by Hooker, was elected a fellow of the Linnean Society. By contrast, Elizabeth Warren was a genteel woman with apparently conventional gendered beliefs about women and modesty. She admitted to William Hooker in the early stages of discussion about her botanical chart that she had "some repugnance . . . to come before the public as an author." [37] Her local scientific community also had conventional ideas about woman's place. She was a founding member of the Royal Cornwall Polytechnic Society in Falmouth in 1833 and submitted many collections of specimens to its annual exhibitions and competitions. But at its annual awards, whereas men received individual public listing for their work, women usually were grouped together, anonymously, as "lady amateurs." Within this climate, although Warren participated actively for thirty years, she "was never able to rise above a place on the Ladies Committee . . . and only became a judge in later years." [38]

Isabella Gifford, one of the female algologists in her network, commemorated Elizabeth Warren's contributions to botany in a memorial sketch in 1865: "When her attention was first turned to this branch [of botany], there were few works on the subject, and the large number of seaweeds found on the British shores were not discriminated and arranged as they are now." She noted how zealously Warren collected indigenous seaweed, encouraged others to join in this area of botanical work, and "was always ready to enter upon any new field of inquiry." [39] Isabella Gifford (ca. 1823–91) herself collected and studied plants of the sea and seacoast, exchanged specimens with those who shared this enthusiasm, and corresponded widely with other botanists. She exchanged specimens and queries with G. A. Walker Arnott, professor of botany at Glasgow, for example, discussing the names and classifications of plants. "Can you tell me," she asked, "anything about the precise locality for Scirpus holoschoenus in Somerset?" "I wish all the cultivators of botanical science afforded information so kindly and ungrudgingly as you do." [40] Born in Swansea, Gifford lived for many years in Falmouth and Minehead, Somerset. Her mother, said to have been "a rarely gifted and most cultured woman," educated her daughter herself. Gifford did not marry, and she remained at home. Her parents fully encouraged her scientific bent, said to have shown itself when she was young. When she died, the obituary in the *Journal of Botany* labeled her "the last link in the chain of lady phycologists," connecting Ellen Hutchins in the early nineteenth century

Isabella Gifford (ca. 1823–91). (Courtesy of Hunt Institute for Botanical Documentation, Carnegie Mellon University, Pittsburgh)

with Mrs. Griffiths, Mrs. Gatty, Miss Ball, Miss Cutler, Elizabeth Andrew Warren, and Emily Stackhouse.[41]

Colonial Travelers

Unlike male botanists who made grand treks in search of new and rare specimens to enlarge empirical knowledge and satisfy the collector's hunger, few early and mid-Victorian women with botanical interests journeyed beyond the local terrain of garden, woods, or seashore. Physical geography and social ideology limited that topographical expression of curiosity. Nevertheless, a few women, notably colonial wives, were able to be "social exploratresses" who traveled to exotic outposts under the aegis of empire.[42] Aristocratic women, daughters of gentry families, and military wives were enabled by their status as wives to journey and so-journ in ways that would have been more difficult at home. From India, Canada, Australia, and South Africa they sent botanical reports, draw-ings, sheets of dried specimens, and boxes of living plants. During the 1820s Lady Amherst, wife of the governor-general of India, had collected plants on her travels, developed a herbarium of Himalayan plants and introduced specimens from the Indian flora into Europe, among them *Clematis montana*. The countess Dalhousie, wife of the commander-in-chief (later governor-general) in the East Indies, 1829–32, also developed a herbarium of indigenous specimens during her stay; at an earlier posting she had collected plants in Nova Scotia. A few decades later Lady Char-lotte Canning, wife of Lord Dalhousie's successor as governor-general of India, collected specimens, visited botanic gardens, traveled with her husband on official tours to remote areas, drew and painted plants, and recorded some botanical findings in a diary. On one long camp journey to the borders of Tibet in 1859, Lady Canning's portfolio of drawings was the only item she saved when her tent caught fire. During a later expedition to Darjeeling, she collected and dried ferns and drew plants; she wrote in a letter: "I have been very busy over my scrap-books of sketches, and writing, or causing to be written, the names, etc. and it is all rather tidily done now—two volumes and a portfolio full, a few more flowers."[43] Marianne Cookson, the daughter of gentry from the north of England and wife of a military officer, also contributed to drawing the Indian flora. In India during 1834, she completed thirty botanical paint-ings of indigenous plants. Her folio-sized, brightly colored depictions of a water lily, lotus, banyan tree, and cashew, detailed in some cases to show the flower or fruit in different stages of development, were published as *Flowers Drawn and Painted after Nature in India* (1835).[44]

Colonial wives were enlisted in projects for imperial botany. William Hooker honored his correspondents by his attentions, gave them status, and validated their interest in supplying specimens for world floras. They sent home requests for equipment to assist their botanical work, and he shipped books, paper, and microscopes out to them. Annabella Telfair, for example, lived in Mauritius, where her husband, a high-ranking colonial officer, was supervisor of the botanic garden. She shared his interest in local flora. At Hooker's request, she collected marine plants. She was not hopeful, though, about the value of her contributions, writing in August 1829: "I shall be very glad to forward you my spoils as they come in, but this is a miserable place for any thing of science or taste." [45] Several of her drawings of plants appeared in the *Botanical Magazine*.

During the 1830s, Mrs. A. W. Walker made many contributions to the flora of Ceylon by collecting and drawing plants. Her husband, a British army colonel stationed in Ceylon, wanted her to "acquire the art of drawing flowers from Nature, Botanically, as he says at Ceylon a valuable collection ought to be made." She corresponded with William Hooker before traveling to her husband's colonial posting and enthusiastically enlisted herself to serve botanical science. In September 1830 she asked Hooker to help her develop the requisite skills. Teachers, she claimed, "never think of anything but the *exterior* of a flower, and *labor* their drawings in so minute and tedious a manner that the plants must . . . be completely dead before the drawing would be half done." She asked to see some of Hooker's own botanical drawings—"they, I am convinced, would show me exactly what I *ought* to do if I *could*." Hooker, eager to cultivate botany in the colonies, met her request. Over the next several years he helped Colonel and Mrs. Walker further their botanical work by sending paper for drying their specimens, and also books. He even filled Mrs. Walker's request for a *"Powerful* Magnifying Glass . . . to use . . . in your service." Mrs. Walker and her husband made several tours in Ceylon, collecting and drawing plants, and sent "many rarities" to Hooker. Several letters detail the often insuperable difficulties of conveying seeds, seed pods, living plants, and dried specimens from Ceylon; her letter of January 4, 1836, reported, for example, that, fulfilling Hooker's request for mosses, she was sending a box of specimens with a friend "who has promised to keep the Box in his Cabin" during the long sea voyage. Mrs. Walker also sent Hooker her account of several of their Ceylonese botanical excursions, which Hooker published in his *Companion to the Botanical Magazine* in order to encourage others to submit accounts of botanical travel to him. "I could wish," he wrote, "that many other spots in our distant colonies . . . might meet with an equally faithful scientific journalist." [46] Two plants were named for her (*Patonia* and *Liparis walkerii*).

Women in outposts of empire botanized for the most part within the gender economies they brought with them from home. Often their botanical work belonged to the culture of female accomplishments, as they collected and pressed plants into albums or made watercolors of flowers. The colonial encounter also made it possible for some women to step outside conventional gender boundaries; in the case of Mrs. Walker, who was clearly trained as a helpmate, gendered commissions gave her larger opportunities to travel and to seek wider fields. But colonial life also generated financial necessities that in turn led women into new fields for action. In Upper Canada, Catharine Parr Traill (1802–99), daughter of a literary English family, faced the exigencies of a settler's life in the backwoods of Ontario and developed polite botanical skills into a livelihood for her family. She collected flowers, observed plants, studied geographical distribution, learned medicinal uses of plants from native women in her district, and produced a series of books over many decades.[47] Women travelers with scientific interests and more secure funding set out for exotic botanical fields to fulfill other ambitions as well. Marianne North (1830–90) was the daughter of an urban widowed gentleman who provided her with art lessons and took her on extended travels in Europe and the Middle East. After his death she traveled to Canada, Jamaica, Brazil, Japan, Australia, the East Indies, and Ceylon, set up her easel, and painted flowers in a sensuous mode that was far removed from either the austere perfection of George Ehret in the eighteenth century or the decorous arrangements of the mid-nineteenth-century Anne Pratt. During the 1870s and 1880s she set out to amass paintings of tropical plants of the world, seen in situ, and became one of the late Victorian "traveling ladies," moving well beyond the breakfast room and the drawing room, and also well beyond the boundaries of polite botany.[48]

EIGHT

Flora's Daughters in Print Culture, 1830–1860

I have not written for the learned naturalist and the stern critic; I have written for the young, the enquiring, and the kind.

Mrs. E. E. Perkins, *The Elements of Botany*, 1837

It is gratifying to me to find that my efforts to simplify this study have been approved of by those conversant with the subject, and that the credit has been awarded me "of having first led attention in a simple, popular, as well as strictly scientific manner, to an interesting branch of botany previously little studied."

Isabella Gifford, *The Marine Botanist*, 3d ed., 1853

By the 1820s botany was so gendered as a feminine area within science culture that some botanists began trying to reclaim it for men. Nevertheless, as botanical science became increasingly sex segregated and male identified over the next decades, literary botany and educational botany remained open fields for women, who could achieve publication more easily through popularizations and educational writing than through more technical books and papers.

Most botanical writing by women during 1830–60 was introductory and elementary and took the form of books for "the general reader" and for women and children. Matching the tenor of Victorian England, some books have a more pious and evangelical tone than was commonplace in women's botany books from the earlier period. Some women wrote "for the love of flowers" and shaped the "language of flowers" tradition in England. Others melded art and science in illustrative works. Books by women included field guides about seaweeds, regional floras, and introductory textbooks, and many authors mirrored the emergent interest in ferns and fungi. Some of their introductory books taught about the natural system, but many others continued to teach about Linnaean botany well into the nineteenth century. In general, women writers during this period continued to perform important work as cultural mediators of botanical knowledge, and their expositions often served as "first steps" to botany.

In contrast to much writing by women about botany during the Linnaean years, 1760–1830, the voices of women botanical writers in early Victorian England were muted and more tentative, less intellectual in range and aspiration. Across the range of women's writing about botany, women authors made fewer claims even to be writing science. Moreover, the narrative forms of women's botany books changed. Earlier, as we have seen, conversations and letters were conventional genres for writing about science. The familiar format taught botany along with general knowledge and often featured a mother-educator who was botanically active at home. Family narratives began to disappear after the 1830s, and by the 1850s the familiar format was erased from women's botanical writing.

Although most Victorian women of the "journeyman" class who wrote for money looked to fiction, some wrote poetry and still others wrote for juvenile readers.[1] Amid the changing climates of early Victorian floral

Eliza Eve Gleadall, *The Beauties of Flora* (1834–37). (Courtesy of Hunt Institute for Botanical Documentation, Carnegie Mellon University, Pittsburgh)

culture, botanical culture, and science culture, popular science writing offered another profitable path to some women, who made gender ideology and cultural codes work to their economic advantage. Sarah Bowdich Lee, for example, wrote *Elements of Natural History* (1844) and *Trees, Plants, and Flowers* (1854), and Agnes Catlow and Maria Catlow produced a variety of books about plants, insects, and shells, including *Popular Field*

Botany (1848) and *Popular Greenhouse Botany* (1857). A few women published as extensions of other activity; Elizabeth Twining delivered lectures to young women who attended classes given at the Working Men's College in London and later issued them as *Short Lectures on Plants for Schools and Adult Classes* (1858).

During the 1830s the fashion for books about the language of flowers offered women writers a way to combine botany, art, and morality. One example, *The Beauties of Flora* (1834–37), consists of forty folio-sized lithographs of flowers "drawn from Nature" and embellished by botanic, poetic, and emblematic material. Each flower in this collection is identified botanically and located within both the Linnaean and natural systems of plant classification, and there are instructions about mixing shades and tints to color the plates. The compiler, Eliza Eve Gleadall, acknowledged the assistance of the proprietor of the illustrated *Botanical Magazine*, and her drawings have a botanical specificity that suggests their use in drawing lessons. A multipurpose book, part drawing book, part emblem book, part compendium of verse, *The Beauties of Flora* is positioned as affording "a chaste recreation" for youth, blending "information with amusement." "There is religion in a flower," Gleadall declares, and hence she also includes information about flowers as emblems; the garden wallflower is "Fidelity in Misfortune" and lily of the valley is "Purity and Return of Happiness." Eliza Gleadall ran a girls' school in Yorkshire, and the number of clergymen on the subscription list for her book suggests that the project accorded with the sensibilities of her audience.[2] Charlotte Elizabeth Tonna, another writer who published within the climate of evangelical piety, presented "floral biographies" in *Chapters on Flowers* (1836). Native plants serve as "momentos" of a person or a sentiment illustrative of "the Christian character." The hawthorn tree leads this evangelical writer to reflect on the "changeableness of earthly things . . . always a favorite and a fruitful theme, alike with the worldly moralist and the more spiritual instructor"; and the jessamine tells a "well-remembered story of patience, piety, and peace." *Chapters on Flowers* is an emblematic and associationist account of plants in the evangelical tradition, with many references to "early happy death," the importance of warring against the kingdom of darkness, and "the Healer who is always nigh."[3] Charlotte Elizabeth Tonna was a militant novelist of social evangelism and editor of the *Christian Lady's Magazine*. She drew on a female preaching tradition within evangelical Christianity that, it has been argued, helped gain women writers a published voice; like Hannah More before her, she came to public authorship through religious discourses and shaped literary power for herself within patriarchy.[4]

Bowdlerizing Botany: Mrs. E. E. Perkins

Even though home-based family narratives of natural history and botany came under attack during the 1820s, the familiar format did not entirely disappear from women's popular science writing. Sarah Waring, author of *A Sketch of the Life of Linnaeus* (1827), used a family framework in *The Meadow Queen, or The Young Botanists* (1836), a book designed "merely for beginners and those of a very youthful age" that combines rudimentary Linnaean botanical information with literary flower associations.[5] Likewise, Louisa Ann Twamley retained a family narrative in *Our Wild Flowers, Familiarly Described* (1839), in which an aunt teaches a niece to give "young flower-loving friends a little pleasant information, without any difficult terms, or unexplained names"; the narrator recalls "what kind of knowledge I should have liked, when I used to wander by hedges and woods and lanes, gathering everything and knowing nothing."[6]

Nevertheless, narrative tides were changing in women's science writing as writers and publishers sought out new expository styles. Mrs. E. E. Perkins's *The Elements of Botany* (1837), one of a number of introductory books by that title issued during the 1830s and 1840s, used no narrative features associated with the familiar format. Directed to young women, it echoes earlier promoters of botany in hoping that the study of botany "may supersede some of the less important, not to say superficial studies, in the education of many of our sex in the present day." But there is no mentorial figure, and the educational mandate is filtered through more explicit religious language than we find in botanical writing of the previous generation. Thus the author declares that botany "lead[s] the mind to contemplate the adorable perfection and goodness of HIM who has bestowed so much of grace and art even upon the inanimate and insensate objects of his creation. No department of Natural History more strongly evinces the existence of a *Supreme Cause* than the study of the vegetable world."[7]

In this period, although "serious" botanists in England were turning their sights toward Continental taxonomies, popular Linnaean handbooks and introductions continued to appear. A review from 1833 of a book about Linnaean terms called it "an elegant and useful manual to ladies and others commencing the study of botany by this system, . . .

who dare not commence the science by the natural system."[8] In her *Elements of Botany* Perkins declared the Linnaean system "the best, from its originality, its distinctness, [and] its easy application."

But Linnaean plant sexuality continued to signal cultural danger, and writers responded to it according to their own beliefs, or the beliefs they imputed to their audience. Introducing *Elements of Botany* as a new elementary botany book for a particular audience, Perkins wrote: "It is true that many works exist of established reputation and considerable merit, but a majority of them are little adapted to the tastes or the capacities of the young; and, from another cause, many are peculiarly unsuited for the perusal of female youth." Her *Elements of Botany* is an example of expurgated botany issued during the period when the Bowdler family was producing sanitized versions of Shakespeare; Mrs. Perkins paraded her own bowdlerization of Linnaeus and explained her exclusion of "all the exceptionable analogies" in this way:

> It is the purpose of this work to present an Introduction to Botanical Science, according to the system of Linnaeus; but care has been taken to divest that system of certain repulsive excrescencies which have operated, not merely to the injury of that system, but even to the detriment of the science itself. This is greatly to be regretted: Botany ought not to be held accountable for the bad taste of its professors.

Writing for young female readers, she determined to expound botany "properly," to vindicate it as "a study perfectly unexceptionable, and . . . harmless" for her chosen audience.[9]

How then, in a bowdlerized book about Linnaean botany, does an author present plant reproduction? Concerning the "organs of fructification," Perkins used the technical terms "pistillum" and "stamen" but omitted references to "husbands" and "wives," "brides" and "bridegrooms." She took a distinctly scientific approach to her exposition and subdivided her topic by describing the "collective organs" that compose the pistil and the physiological contributions of each to "the business of perpetuating the species." She also criticized Linnaeus's belief in "resemblance between the vegetable and animal creation" and, like Agnes Ibbetson, judged animal analogy to be a hypothesis not borne out by microscopic observations. About the analogies central to Linnaeus's sexual system she wrote: "The ideas which they represent are in no respect essential to his system, but in truth have the effect of excrescencies and deformities; by introducing vague and fanciful analogies they detract from the severe truth and propriety by which it is in general characterized." [10]

Mrs. Perkins was an enterprising promoter of botany and of her own writing and drawing. One novel feature of her book was its material connection between botanical study and proper botanical instruments. "The Authoress . . . has arranged a model from which cases of instruments are in progress of being made, adapted for the boudoir, the garden, or the fields, by which the allurement to a minute knowledge of the subjects of botanical science may be materially facilitated." She linked her book to a course of instruction available through her publisher. "Each purchaser will be entitled to a Card of Admission to one Lecture on the uses and the mode of applying the several instruments, which will be delivered at the Publisher's, every Wednesday and Thursday morning, from twelve to two." She was also a self-described "Professor of Botanical Flower Painting" and probably either had a school or taught flower painting from her home. Perkins explained in the preface that she originally had intended to publish a purely illustrative work titled "The Boudoir Recreations in Botany" but turned instead to the expository task in *Elements*. An "advertisement" announces her plan to "continue the subject in its higher departments" by publishing an illustrated book titled "Physiological Researches in Botany." She later completed illustrations for James H. Fennell's *Drawing-Room Botany* (1840), a short book about Linnaean plant classification that shows details of the parts of flowers. [11]

"For Popular Use and General Interest": Anne Pratt

Anne Pratt (1806–93) was one of the best-known women writers of "popular" botany texts. In 1849 the critic Thomas Macaulay defined a popular style in writing as one "which boys and women could comprehend," [12] and Pratt's many books, written over the course of a long career, addressed these groups of readers. She wrote especially for young students—always denominated "he," but hers were not school texts for formal classroom work. Rather, her books are descriptive botany and generalist accounts of plant folklore and plant uses. They range like a panorama of the vegetable kingdom, from wildflowers to ferns to grasses to poisonous plants and seaside plants. Pratt positioned herself at the lowest level of introductory and elementary work, standing at the gateway to science, helping others to pass in, like a botanical Virgil guiding pilgrims to the promised land of botanical science that she herself could not enter, or would not enter (or strategically, as a woman writer, should

Anne Pratt (1806–93). (Courtesy of Hunt Institute for Botanical Documentation, Carnegie Mellon University, Pittsburgh)

not enter). Pratt wrote her books at a moment in the history of botanical culture when the codes for asserting scientific authority were hardening and when channels for women's writing were narrowing. She also wrote for money at a time when interest was high in diffusing knowledge to new bodies of readers. Her way was to blend botanical information with the romance of nature, and by that means she did important work for a community of nonscientific general readers.

During the 1840s and 1850s, when the fashion for exotics grew to fever pitch, Anne Pratt wrote about native plants and wildflowers. Her books celebrate an older folk tradition by introducing indigenous flowers as repositories of local folk customs and remedies. Her best-known work was the five-volume *Flowering Plants and Ferns of Great Britain* (1855). Adapted "to the use of the unscientific," it guides the reader through each of the natural orders of plants and describes British examples. Pratt used English terms throughout, "to aid those who have not hitherto studied botany," and allied herself with the tradition of generalist and popular writing by including many references to the uses of plants and by presenting historical details and folkloric information (e.g., about the "tonic properties" of sea holly (*Eryngium*), whose "candied roots were introduced into general use by Robert Buxton, an apothecary; . . . the town of Colchester was long famous for this sweetmeat"). In that decade she also wrote about ferns and grasses. *The Green Fields and Their Grasses* (1852) is an illustrated informal guide to help young readers and general readers identify common species of grasses "such as make our meadows, and hills, and valleys green by their abundance." *The Ferns of Great Britain, and Their Allies the Club Mosses, Pepperworts and Horsetails* (1855) is a nontechnical, yet not overly simplified, account of native ferns for "persons who are fond of nature, and who have yet never studied Botany systematically." [13]

Anne Pratt's most widely reprinted work was *Wild Flowers* (1852–53), a little two-volume book for children illustrated with block prints. It opens this way: "Every child who has wandered in the woods in the sweet months of April and May, knows the Blue-Bell, or Wild Hyacinth." A few descriptive details accompany each flower, with comments about identification and use. When writing about "Corn Feverfew (Pyrethrum inodorum)," for example, she explains that the young botanist can often be perplexed by similarities in general appearance between it, some species of camomile, and the tall oxe-eye daisy. "This flower has little claim to its familiar name of May Weed, for it blossoms from August until October, on fields and waste places, and is among the few flowers which we may perchance find even in the very depths of winter." She continues: "Its stem is about a foot high, and it has a slightly aromatic odour. . . . The Common Feverfew (Pyrethrum parthenium) . . . [is] one of the plants which among country people are held in high estimation as a remedy for fever." [14] *Wild Flowers* found favor with Queen Victoria. Pratt had written in September 1852 requesting permission to dedicate the work to her and thereafter was instructed to send copies of all her future works to the palace for Queen Victoria's children. The publisher also issued the book in single sheets, to be hung in schoolrooms.

Anne Pratt's personal story replicates the profile of other women in botanical culture who came to botany through family networks and were able to pursue it because it was congruent with social norms. Her father was a prosperous wholesale grocer in Kent, and her mother descended from flower-loving Huguenots. She was a delicate child, lame from an early age, and a male family friend introduced her to botany as appropriate for an intelligent girl with physical limitations. Botany became her favorite avocation. Her older sister was her botanical assistant and companion, searching out specimens of plants for her. (As botanical sisters, they enacted the narrative of introductory botany books in the familiar format during their day.) Pratt compiled a herbarium and also cultivated her artistic skills by drawing the plants in her collection.[15]

When she was thirty-two years old, Anne Pratt published her first book, *The Field, the Garden, and the Woodland, or Interesting Facts Respecting Flowers and Plants in General* (1838). She had kept the work secret from her mother and friends, and the book appeared anonymously. After her mother's death in 1845, she lived with friends. It is unclear what provision her family made for their lame and unmarried daughter. Books seem to have been a source of income. At age sixty she married a man she had known in earlier years and produced new editions of the botanical works that established her reputation. Pratt's publications included natural history books, religious commentary, and a collective biography of eminent eighteenth-century figures, including Sir Humphry Davy, Cuvier, and Madame de Genlis, "to convince the young of the importance of cultivating both the mind and the heart."[16] She published her early books with Charles Knight, publisher of the *Penny Magazine*, and then with the evangelical Religious Tract Society. Thereafter her main publishing base was the Society for Promoting Christian Knowledge. Her many SPCK books used their religious brush lightly but used it, and Pratt's kind of popular botany writing can be read as illustrating the "safe science" that educators and religious publishers were promulgating during the 1830s and thereafter.[17] *Poisonous, Noxious, and Suspected Plants of Our Fields and Woods* (1857), for example, was issued in several formats, including a large-print cheap edition for working-class readers.

In an early book, *Flowers and Their Associations* (1840), Anne Pratt articulated the audience that she continued to address in all her subsequent writing about plants; she wrote for "the general reader," one who is "fond of flowers" but has not made them "the object of their study." Her writing is sweet, even-tempered, and unpretentious. *Chapters on the*

Common Things of the Sea-Side (1850) opens this way: "It is delightful on some fine summer's morning to wake up to the loud continuous sounds of the waves, and to stray along the shore, with eye and heart alive to the natural beauty of this world." This natural history book, with sections on marine plants, shells, and marine animals, melds observation, aesthetics, and emotion into a romance of nature similar to writings by Gosse and Kingsley. She aims to "awaken" an interest in nature among "the unlearned" who find themselves resident at the seashore for reasons of health, and therefore she touches briefly on hundreds of subjects. Pausing for a moment over upright sea lyme-grass, for example, she describes its size and appearance and then comments on its uses: "A coarse sort of fabric was formerly made of its leaves; hence its botanic name of Elymus is taken from a Greek word, signifying to cover." She gives some room in her book to poems about plants but is more interested in culinary uses; sea purslane, she notes, "makes a good pickle." [18]

Two biographical accounts construct her career in ways that serve different communities of interests. In 1889 the *Women's Penny Paper* carried an interview with the very elderly Anne Pratt, conducted by her niece. This important feminist weekly newspaper ("The Only Paper Conducted, Written, Printed and Published by Women") covered women's organizations, focused on industrial, social, and educational questions, and profiled leading women of the day. In the tradition of exemplary biography, the article praises her work and her character, surveys her writing, and refers to her "genial, poetic, sympathetic, yet scientifically accurate temperament, which has always charmed her readers." Pratt, we read, "enjoys a popularity among all lovers of Wild Flowers, that has made her name a household word in many homes; and being one of the earliest women botanists she has perhaps done more to popularise the study of Botany than any other woman writer." [19] The article yokes botany and happiness: "There is little doubt that the study of Botany, like that of all Natural Sciences, gives dignity and elevation to the character, and certainly in the classification of these 'relics of Eden's bowers,' Anne Pratt has found what Keble says so few discover, 'the happy secret of their calm loveliness,' for when visiting her a few weeks ago she remarked, at the advanced age of 82, her countenance beaming with gratitude, 'I have had a very happy life.'" The biographical sketch in the *Women's Penny Paper* incorporates languages of nature that promote science for women by combining literature, home, and moral benefit.

By contrast, a memorial account of Anne Pratt in the *Journal of Botany*, while appreciative of her many contributions, is clear about her location on the margins of scientific botany. James Britten, a biographer of British botanists, wrote that "in the strict sense of the word, Miss Pratt was not a botanist: the introductory portion of her principal work [*The Flowering Plants and Ferns of Great Britain*] is singularly meagre and incomplete, and her descriptions show no intimate knowledge of the plants described. She was, however, a true plant-lover, and by her pleasant style succeeded in interesting her readers in the objects she described."[20] Britten likewise reported meeting the elderly Pratt. He found "a bright intelligent little woman, full of interest in natural history subjects, a certain kindness and simplicity of manner, coupled with much gentleness" who reminded him, he wrote, of the character Miss Mattie in Elizabeth Gaskell's *Cranford* (1853). In that novel, contemporaneous in composition with Anne Pratt's books, the kind, uncomplaining, and comforting Miss Mattie lives by the rules in her rural English village and is courageous in the face of adversity but not contestatory. James Britten's Mattie-like Anne Pratt is intelligent and conscientious, writing sympathetically for children and general readers, not chafing at her station, encouraging but not haranguing. Hers is a voice of literary botany, one that knows its place.

Compared with earlier Enlightenment-style narratives by women botanical writers, Anne Pratt's books are modest, make few large claims, and promise no dramatic transformation through natural history study. Unlike Priscilla Wakefield with her belief in the transformational possibilities of botany for girls, Pratt held out no explicit educational hand to female readers. Whereas books in the familiar format establish authorizing and enabling narratives of expert women teachers at home, Anne Pratt's books supply no equivalent narrative for investing her texts with authority. She benefited, nevertheless, from the cultural alignment of writing women and religiomoral publications. By choosing to avoid botanical terminology, by eschewing theoretical discussions about classification, she did important work for her general readers. Like many other Victorian women writers, Pratt sets challenges for feminist biography. She did not mask her own name on her title pages, yet the books themselves seamlessly obscure the author and supply little information about the circumstances surrounding their publication. Some or all of this may have been a successful strategy by a woman author in early Victorian England, threatening no one, according with gender ideology, shaping a narrative voice that brought her contracts with the religious publishing industry. An important popularizer of botany, she knew how to shape and work the codes of early Victorian literary and botanical culture in order to carve out authorial space for herself.

Pictorial Floras: Anna Maria Hussey

Within the visual area of Victorian print culture, some women turned the polite skill of botanical drawing to their own educational and economic purposes. Illustrated botanical books were one way to meld art and science. Mary Anne Jackson, for example, produced *The Pictorial Flora* (1840) as a manual of illustrations of British plants and an adjunct to the main descriptive works of her day. This book consists of fifteen hundred small lithographs of flowering plants indigenous to Great Britain. One index of Latin names and another of English names assist in plant identification. The plates in this book are arranged so that they could be bound into Smith's *English Botany* or works of descriptive botany by other authors. Jackson explained in the preface that books of illustrative plates can assist in examining plants when no scientific instructor is available. Because many books of plates are "too unwieldy for pocket-companions" and are "so expensive as to preclude their being put into the hands of very young persons," she chose the path of small drawings and designed the book for cheapness and portability. Jackson positioned her work as a "pioneering task" and as a trial run for a larger project (which never materialized) "to bring before the public a cheap and complete Pictorial Encyclopedia of British Botany."[21] Elizabeth Twining's *Illustrations of the Natural Orders of Plants* (1849–55) is an ambitious two-volume folio publication that contains 160 color plates of indigenous plants. Rather than a study of minute scientific detail, this pictorial publication, produced as "a guide into the wondrous garden of Nature," featured plants as "the gifts of a heavenly Father, accessible to all His children."[22] Other pictorial floras brought exotic plants home to England from the colonies. Arabella Roupell, born into a well-to-do clergyman's family in Shropshire, married a clergyman who was a member of the East India Company and lived in India and then in South Africa for some years. During the 1840s she drew plants collected at the Cape of Good Hope; she drew them, by her own account, "solely for the amusement of leisure hours." But after the visiting botanist Nathaniel Wallich carried a selection of her paintings back to England and showed them to William Hooker, they were issued in a volume titled *Specimens of the Flora of South Africa by a Lady* (1849), accompanied by a descriptive text.[23]

During 1830–60 the art of botanical illustration was in rapid flux. New

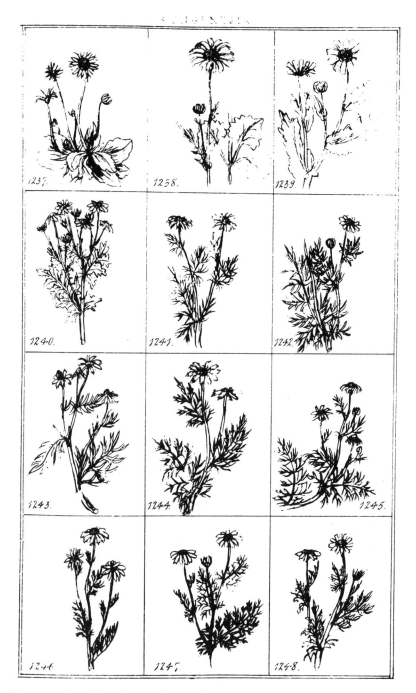

Illustration from Mary Anne Jackson, *The Pictorial Flora* (1840).

media were developed, including lithography, photography on paper, and nature printing. Anna Children Atkins, a pioneer in photographic botanical illustration, used the newest technology to catalog specimens. She was "the earliest woman photographer and also the first person to print and publish a photographically illustrated book."[24] A few years after William Henry Fox Talbot invented photography, and before he published his earliest photographic books, Anna Atkins applied the cyanotype process to botany. In this chemical process, involving neither a camera nor a negative, specimens are placed directly on light-sensitive paper; the result, when developed, is an image familiar to us now as the architectural blueprint. Using the cyanotype process, Anna Atkins produced thousands of "sun pictures" or "photograms" of seaweeds, and later of ferns. Her privately printed *Photographs of British Algae: Cyanotype Impressions* (1843–53) consists of approximately four hundred handmade photographic prints and was meant as an illustrative companion to William Harvey's botanical reference work *Manual of British Algae* (1841). Anna Atkins applied the new technology to compiling images for scientific study and set out to record marine plants and help botanists distinguish one species from another.

During the 1830s, 1840s, and 1850s, cryptogamic botany attracted interest from specialists and novices, who collected fungi, mosses, and lichens, studied them with microscopes, and recorded them visually in watercolors and paintings.[25] Anna Maria Hussey's *Illustrations of British Mycology* (1847–49, 1855) attests to this new world of interest. It is an expensive two-volume quarto series of color plates of fungi, morels, truffles, and puffballs, accompanied by botanical information and general commentary. *Illustrations of British Mycology* is avowedly a popular work rather than one written for specialists and a book for wealthy readers rather than an inexpensive field guide for workers. In presentation and price meant for the drawing-room table, it is generically borderline, a hybrid of fashion and science, and straddles the ornamental and the informative.[26] A reviewer in the specialist *Journal of Botany* who praised the "artistic merit" of Hussey's work, along with her "considerable care giving details," designated it as "more suggestive of the drawing-room than the study."[27] But this illustrative book goes beyond the conventionally fashionable and suggests how serious early Victorian botanical hobbyists were in their popular science work.

Illustrations of British Mycology presents portraits of mushrooms in the order Hymenomycetes, a taxonomic grouping of fungi whose spore-

Anna Maria Hussey's *Agaricus squarrosus*, from her *Illustrations of British Mycology*, ser. 1 (1847–49), plate 8. (Wellcome Institute Library, London)

bearing surface is exposed; this order includes the families of agarics and polyporus. For each specimen, such as *Agaricus squarrosus*, Hussey cited lengthy botanical descriptions of the genus and subgenus of each fungi and identified habitats, seasons of growth, and locations. A prefatory essay introduces terminology and describes the orders of fungi. Her commentary to the 140 illustrative plates blends science, anecdote, and literary reference. There are frequent first-person references, often with a literary flair; thus, "I have mourned over specimens nearly a foot across, their pure ivory gills and glowing scarlet pileus crushed in the dusty road." Discussion of the utility of fungi incorporates a recipe for omelette of Giant puffball. Hussey also promoted wider moral and educational reasons for studying fungi; even those specimens with no known utility are useful if they "induce us to study Nature's works." Writing with girls and their mothers in mind, she remarks on those without interest in nature, who take "the languid duty walk" and show "apathy toward plants and simple objects in the path." Blending scientific knowledge and moral betterment, she offers advice and social analysis: "In whatever district they reside, some external pursuit . . . will present itself, and habits of observation and amusing tastes may be formed, useful through life for the same ends. But if Mama object to the little fingers which present her a Fungus, that it has soiled the glove; . . . if servants are allowed . . . to scold at stepping across a ditch for the flower: Mama must not in reason complain if the Young Lady be . . . vapid . . . and, in after years, unamusable, delicate, repining . . . instead of finding occupation and a charm in everything." [28]

Her book was meant, then, for general readers and for nonspecialist purposes. To help awaken readers, including students and the parents of potential students, to the pleasures of mushrooms and other fungi, she also described the equipment needed to collect and analyze them, foremost among them a proper basket for collecting specimens. Her spirited interjections are part of the charm of her book. She warns that "if the student should leave home without [a basket], a profusion of lovely and rare objects will be certain to strew his path." If one forgets the proper basket for collecting fungi one has, she writes, "but two alternatives, to dissect on the spot, always an imperfect operation, or to carry away the spoil in hat or handkerchief, when on arrival at home, a heterogenous mass of caps, stems, etc. presents itself—*disjecta membra*! who shall assign to each its proper parts?" [29] Hussey's pleasure in searching for fungi is serious but not ponderous and appears to have been based on keen interest in plant study. Her quest as a student differed from the motivation that led Charles Darwin's daughter Henrietta to invent the "sport" of hunting out and eradicating a fungus that offended her, namely, the *Phallus impudicus*, the phallic-shaped fungus nicknamed stinkhorn because of its odor. Her determination was documented in a family memoir: "Armed with a basket and a pointed stick, and wearing a special hunting cloak and gloves, she would sniff her way round the wood, pausing here and there, her nostrils twitching, when she caught a whiff of her prey; then at last, with a deadly pounce, she would fall upon her victim, and poke his putrid carcase into her basket. At the end of the day's sport, the catch was brought back and burnt in the deepest secrecy on the drawing-room fire, with the door locked; *because of the morals of the maids*." [30]

Illustrations of British Mycology, in its art and in the vigor of the commentary, bespeaks a keen student of the science. Anna Maria Hussey (1805–77?), an artist and writer, was also a Victorian clerical wife who lived in the country and collected fungi near her home in Kent. During the 1840s she corresponded with Miles Berkeley, a Northamptonshire clergyman-naturalist and well-known mycologist, who wrote the section on British fungi for William Hooker's edition of Sir James E. Smith's *English Flora* (1836). She wrote to him as a student to an expert, reported on her findings and observations, enclosed specimens of fungi, and asked his assistance in identifying and naming those that did not match published descriptions (perhaps including those in his own book). She also offered to serve as his mycological assistant. Letters show her immersed in drawing specimens when they are in their prime: "I have been painting Fungi 8 hours today —!" and "I am so tired of collecting and painting them — I dream of them all night, as well as work all day." [31] In a lighter vein, she reported culinary successes and failures with cooking fungi, such as which agarics were particularly suitable for making ketchup and how the "lovely rose blush" upon the stems of one specimen turned to a "very venomous" looking green when cooked.

Hussey's work as a keen collector and botanical illustrator took place in the context of her life as a wife, mother of three children, student, and writer. At one point, although her household was sick with chickenpox and there was no cook, she still managed to send specimens to Berkeley. Her letters display an indefatigable spirit struggling with the manifold responsibilities of her life, including the duties of a clergyman's wife. One undated letter from the 1840s reads as follows: "I thank God most heartily for very good health — rheumatism and ague often attack, but never make me so ill, that a *resolute will* cannot make me capable of exertion. I am however beside my legitimate duties as a mother and mistress — the gardener, the farmer, and just now upholsterer — trying to make old chair covers and carpets passable." Lest Berkeley, himself a clergyman, think she is complaining, she reassures him that hers is a "very cheerful spirit." Nevertheless, she declares herself envious of his life with its clear professional demands. She continues: "Without sufficient occupation I should be miserable — You are better off than I am in one respect — you have regular routine business for good part of your time. Now as I have not, everyone thinks my time at their mercy — morning visitors spoil day after day — every old woman in the parish would complain if I missed a particular hour for an audience." She feels torn among competing demands: "There is not a letter or anything else my husband writes that I am not to read — and to *help him* do everything — I am always under the uncomfortable impression that *the thing I am* doing is not the *right thing to be doing* . . . I never set my palette for oils that some bewitching rarity of the Fungus tribe does not come to me, to say, 'if you don't take my portrait now you may never see me again.' I am never starting in proper costume for a collecting expedition, that D. H. does not remind me how many visits I owe, to people for whom I must put on clean gloves and a good gown." The tenor of this long letter shows Hussey complaining yet labeling her complaints "minor miseries." Nevertheless, she is clearly restive in the face of gendered pressures, particularly those on the clergyman's wife.

Berkeley represents the world of Hussey's choosing, her alternative sphere. There is no evidence in the correspondence that Hussey and Berkeley ever met, yet he was her companion of choice in activities she treasured. Another undated letter, encouraging him to visit, describes the countryside near her home, where there are "woods full of peat and springs . . . thousands of acres of heath"; she promises that "we can carry dirty baskets and tools—and wear old Cloaks and bonnets—and do just as we like." Her fantasy was not so much a fantasy of equality as one of acceptance by a man she called "my most kind and patient tutor in the science." She looked to him to validate her mycological interests and declared herself "charmed" when Berkeley admits his own puzzlement about naming and classifying some specimens of fungi "because every now and then I begin to weary of enigmas—and feel sure Mycology was never intended for women, who love jumping to conclusions better than such patient research."[32] Berkeley showed his regard for her by naming the fungi *Husseia* in her honor.[33]

Hussey's letters to Berkeley date from the years when she was completing illustrations and commentary for her mycology book. She asked for editorial advice about the arrangement of materials in her text. He, responding to her worries, tells her that she may "borrow verbatim the generic char. from the Flora." She, concerned about style levels and readership, tells him that she "always make[s] Dr. Hussey read if it is necessary to suit the comprehension of those not botanists." The first series of the book was published by subscription, and at a high price; she wanted to put Berkeley's name on the subscription list and offered to pay for the volume as her gift to him. Problems attended the illustrations. Expressing "intense disgust" at the quality of the plates for the book, she complained that "the artist has no feeling of his subject at all."[34] Her complaints about the poor quality of the lithography led to troubles with her publisher, and she declared to Berkeley that she would not go on with a second series of illustrations under the auspices of her current publisher, Lovell Reeve. Her own wish was to daguerreotype all the fungi. Nevertheless, she completed drawings for a second series that Reeve published, not by subscription this time, but "taking the risks and profits."

How did Anna Hussey come to her interest in fungi? We have seen that a family pattern of botanizing exists in the genealogies of most women who gave public voice to their botanical interests. Relationships gave access when public institutions did not. Women in clerical families often benefited from religious injunctions to study nature. Networks were an important part of Hussey's story as well. The daughter of a clergyman, Anna Maria Reed married a clergyman whose extended family counted botanists among its members.[35] Her sister, Fanny Reed, shared her interest in drawing and in fungi, and together they completed numerous drawings of English fungi before writing to Berkeley. Through him she became acquainted with Charles Badham, another physician, who, perhaps in retirement, became a curate, turned to mycological work, and wrote a book about edible fungi (*A Treatise on the Esculent Funguses of England*, 1847). Hussey located her project in relation to Badham and to Berkeley's section of *English Flora*: "Dr. Badham having already written a *popular* work on Agarics—and I and my sister having so many English drawings—we are anxious to unite our forces and to get an illustrated work of that kind published. It will be offered to Longmans. . . . We have much encouragement for our plan—ours being intended for popular use and general interest cannot hurt, but will possibly increase the sale of your continuation of Smith. *That* will of course be our principal authority." Allusions in the correspondence suggest that money was not an incidental factor in *Illustrations of British Mycology*. The rector of Hayes came from the Kent gentry, but the family house was small and a son was away at school in Leiden. In 1849 Hussey published a story anonymously in *Fraser's Magazine* and remarked to Berkeley: "This *pays well*, much better than Mycology!"[36]

Anna Hussey came to mycological work from the artistic end rather than from systematics, the field tradition, or study based in microscopy. In another undated letter from the early days of her project she wrote: "I was asked today to state how much I know of Mycology and could only say very little—but I hope to fulfill any promise I make—I am not the sort of person to venture beyond my depth I hope—at any rate I can hold out a hand to guide in shallow waters if I cannot venture into the swell of the Atlantic. I *can* paint—that I am sure of." Hussey neither sought nor had a career as mycologist, artist, or writer. Her engagement in mycology, while clearly fervent, was shaped by the circumstances of her life as a wife and mother during the 1840s. A clergyman-husband may have impeded her work but did not fundamentally hinder it, and her children were helpers; she reported young daughters' bringing her specimens and announcing: "O Mama we have brought you some funny white toadstools."[37] Home was where she worked, and botany was part of her life, but she positioned herself and her work in a way that accorded, we may assume, with both gender ideology and family sexual politics.

A Leicestershire Flora's Nonconversation: Mary Kirby

By midcentury most women botanical writers mediated their knowledge through introductory books. Mary Kirby, the compiler of *A Flora of Leicestershire* (1850), was an exception to this pattern. Although a few publications had appeared before about local plants in that county, hers was the first complete flora of Leicestershire. Her book grouped over nine hundred local flowering plants and ferns according to the orders of the natural system and provided details about habitats and localities. Kirby drew on the work of various clergyman-naturalists.[38] A few preliminary copies were printed in 1848 with alternate pages blank and forwarded to local botanists, and information about fresh species or localities of indigenous plants was in turn integrated into what became the full edition of *A Flora of Leicestershire*. This definitive issue carried substantially increased information about fresh species or localities of indigenous plants and identified finders. Plants were organized according to their natural classes, with "Dicotyledonous or Exogenous Plants" followed by "Monocotyledonous or Endogenous Plants." Notes of general and medical interest appeared in this edition, prepared by Mary Kirby's sister, Elizabeth Kirby.

Kirby's path into science work and writing traces a pattern familiar among women in science culture at this time. From a prosperous hosiery manufacturing family in Leicester that was attentive to her formal and informal education, Mary Kirby (1817–93) studied at day schools and had lessons from various masters. Her father brought home magazines issued by Charles Knight and the Society for the Diffusion of Useful Knowledge that aimed to provide wholesome and useful information.[39] The president of the Mechanic's Institute in Leicester, a family friend, brought her books, taught her languages, and gave her contact with the itinerant lecturers who came during the winter months to talk on scientific subjects; she recalled attending lectures on astronomy, geology, and acoustics, for example. During the late 1830s Kirby was already "deep in the study of

A

FLORA OF LEICESTERSHIRE;

COMPRISING

𝕿𝖍𝖊 𝕱𝖑𝖔𝖜𝖊𝖗𝖎𝖓𝖌 𝕻𝖑𝖆𝖓𝖙𝖘, 𝖆𝖓𝖉 𝖙𝖍𝖊 𝕱𝖊𝖗𝖓𝖘 𝕴𝖓𝖉𝖎𝖌𝖊𝖓𝖔𝖚𝖘 𝖙𝖔 𝖙𝖍𝖊 𝕮𝖔𝖚𝖓𝖙𝖞,

ARRANGED ON THE NATURAL SYSTEM;

BY

MARY KIRBY.

WITH NOTES BY HER SISTER.

"You may run from major to minor, and through a thousand changes, so long as you fall into the subject at last, and bring back the ear to the right key at the close."

LONDON:
HAMILTON, ADAMS, AND CO. PATERNOSTER ROW.
LEICESTER: J. S. CROSSLEY.

MDCCCL.

Mary Kirby, *A Flora of Leicestershire* (1850).

botany." She "ransacked" the beach during a family seaside holiday in Ramsgate, "searching for flowers and grasses and such like, in every hole and corner" and pressing plants under her mattress. She visited the Royal Botanic Gardens at Kew and met Sir William Hooker, "whose British Flora had been my daily companion."[40] One of her female cousins also botanized, having been much affected by the "botanical mania." During the late 1840s Kirby submitted brief botanical notes to the *Phytologist*, a journal for field botanists.[41]

Mary Kirby's work on *A Flora of Leicestershire*, as well as her subsequent writing, had an economic spur. In 1848 her father died; he had suffered business losses during the late 1840s and worried that his daughters, "after all the care and attention he had bestowed upon them, . . . might have to battle in the world." She and her unmarried sisters took on tasks to ensure their future well-being; Mary and Elizabeth Kirby "soon began to plot and plan for book-writing."[42] *A Flora of Leicestershire* was a project toward self-sufficiency, but its publication did not lead her into a career as a writer of mainstream technical botany books. Instead, the sisters fashioned themselves into a professional writing team and wrote more than twenty books for young people.[43] A botanical emphasis nevertheless remained consistently strong in their writing. *Plants of the Land and Water* (1857), for example, a book for young readers, blends botanical information and "curious facts" about the uses of plants in different parts of the world. Written in the form of "Short and Entertaining Chapters on the Vegetable World," it resembles Anne Pratt's books from the same period. The Kirby sisters had suggested this book to the publishers of a new series of illustrated juvenile natural history books; the book, Mary Kirby wrote, was "very easy to do, for I had the botanical knowledge at my finger ends, and Elizabeth had fluency in writing."[44]

Mary Kirby and Elizabeth Kirby identified work compatible with conventions of their gender and class. They worked at home and wrote for children as gender-appropriate professional writers. Mary Kirby harnessed her botanical interests to their economic advantage, and her autobiography reflects the personal and professional rewards writing brought her, including her pleasure in earning money. "Earned money," she wrote, "seems always the sweetest and best of any; . . . we were glad to have found a ready sale for our manuscripts, and also to put the profits into our pockets." Their success was great enough, and regular enough, that they never needed to apply to the Royal Literary Fund. She and her sister Elizabeth were much in demand by publishers, and deadlines often pressed upon them. Mary Kirby recalled a September evening when her sisters went to a party but she stayed home, "quietly pasting down botanical specimens in the breakfast-room upstairs."[45] In her middle years, when Mary Kirby married the clergyman Henry Gregg, she and her sister had money enough to purchase a living for him.

Local family circumstances helped shape Mary Kirby's career as a popular writer on botany and natural history, but her story echoes more widely among women in botanical culture during the 1850s. After *A Flora of Leicestershire* was published, she received an "agreeable" letter from Sir William Hooker and noted "the sympathy [Hooker] expressed when any local effort was made to interest the public in the study of plants, or to make that study popular." (William Hooker had assisted Elizabeth Warren's efforts to promote her *Botanical Chart for Schools* in the late 1830s.) But Hooker's kindness and support for popularizing botany were outweighed by scientific attitudes at that time. Botanical science, embodied in John Lindley, did not make Mary Kirby feel welcome. In a telling anecdote, she recalled in her autobiography spending a day with John Lindley in 1857, seven years after publication of her *Flora of Leicestershire*. They met socially in Norwich, where Lindley and his wife and children had taken a summer house. Mary and Elizabeth Kirby had recently finished writing *Plants of the Land and Water* and, she remarks, "it was well for us, we had, for the great doctor's influence was not calculated to encourage any work, except like his own, of the most scientific kind." The day was rainy, and the party was forced to play indoor games. No botanical conversations developed. There was not, she reports, "much sympathy. . . . I suppose there was no 'elective affinity' between the great doctor and ourselves; for we felt constrained, and not at our ease in his presence."[46]

Mary Kirby's nonconversation with John Lindley in 1857 illustrates the gulf that had opened by midcentury between the specialists and the generalists, the academics and the popularizing writers, those seeing themselves as "scientific" and those marginalized as "more literary." These gulfs in discourses of nature became gulfs in access. In Mary Kirby's case botany was partly private avocation and largely a tool of her public career as a writer. There clearly were thresholds she chose not to cross. In her autobiography she referred in passing to a "humorous" account about the future she and her sister were writing: "of what would happen a hundred years hence; how the men would be thrust out from all the professions by the women, and even the government of the country would be carried on by women; and in the Houses of Parliament there would not be a man to be seen."[47] Kirby made the necessary pragmatic choices within gendered circumstances of her day. Since the path toward botanical science was becoming increasingly male and increasingly professional, she took the less complicated, more welcoming, and also more lucrative path into popular writing.

Botany for Ladies *or* Modern Botany? *Jane Loudon*

By the 1840s several different communities had consolidated within botanical culture. Jane Loudon (1807–58), one of the best known professional writers of popular books about plants during the 1840s and 1850s, produced a steady stream of publications directed principally to women and young readers. Cultivating the publishing markets of her day, she was particularly known as the author of such gardening books as *The Ladies' Companion to the Flower Garden* (1841; 9th ed., 1879), and *The Ladies' Flower-Garden of Ornamental Perennials* (1843–44). Loudon contributed many essays to the *Gardener's Magazine*, the horticultural journal edited by her husband, John Claudius Loudon, and founded and edited the *Ladies' Magazine of Gardening* (1842). Her portfolio also included popular and educational books specifically about systematic botany.[48]

Jane Loudon came into botanical and horticultural writing through her marriage to one of the best-known writer-agriculturalists of the mid-nineteenth century. When the twenty-three-year-old Jane Webb married the much older John Claudius Loudon — industrious and prolific encyclopedist, journalist, town planner, landscape gardener, and writer interested in agricultural and technological improvements — she was already a published poet and novelist. They met when he enthusiastically reviewed her novel *The Mummy* (1827), a work of science fiction set in London in the year 2126, filled with technological projects including tractor-drawn plows, suspension bridges, and milking machines. During their thirteen years together, Jane Loudon was secretary and assistant to her husband and accompanied him on horticultural research excursions around England and Scotland.[49] She had no knowledge of plant science when they met; indeed, she had an antipathy toward botany from childhood because Linnaean systematics bored her. Her botanical education came on the job. Embarrassed that Mr. Loudon's wife could not identify a plant, she began studying botany on her own and attended lectures given by John Lindley. She progressed from embarrassment to study, and then from private botanical knowledge to public writing about plants.

Amid the limiting circumstances of Victorian gender ideologies, Jane Loudon capitalized on her opportunities by writing books about both gardening and botany. Whereas her gardening books highlighted the horticultural applications of knowledge about plants, her botanical books presented systematic botany as a value in itself. *The First Book of Botany* (1841), designed "for schools and young persons," introduced terms common to all systems of botany as a useful first step toward knowledge of plant classification and identification. Loudon explained the parts of plants and types of plants (e.g., herbaceous, succulent), in order to help students study either the Linnaean system or the natural system. In *British Wild Flowers* (1844) she presented botanical information arranged systematically according to the natural system. To illustrate each family of plants, she drew on examples of British wildflowers but, unlike Anne Pratt's books on British wildflowers, hers included no folklore and no information about the uses of plants. Her book presents botanical information, systematically arranged and presented, following Lindley's *Synopsis of the British Flora*, a book "no botanical student should be without."

Jane Loudon's introductory botany books and essays addressed where women were in terms of access to botany. Like Anne Pratt and other professional writers of popular botany and natural history books, Jane Loudon situated herself on the threshold of knowledge, helping her readers cross over into scientific study. Through her writing she campaigned to bring more women into botanical study, aspiring, as she put it in *British Wild Flowers*, to "induc[e] such of my readers as may be unacquainted with botany to study a charming science, which has hitherto been too much neglected." She continues: "I must confess nothing would give me more pleasure than to see botany commonly taught in girls' schools, as French and music are at present. . . . I sincerely hope the time may arrive, though probably I shall not live to see it, when a knowledge of botany will be considered indispensable to every well-educated person." The turn away from the Linnaean sexual system to the natural system helped her project. Although one can argue that the simplicity of the Linnaean system of identification and classification made botanical work easier for beginners, the sexual focus of Linnaeus's arrangement of plants continued to be an impediment to teaching botany to girls. Jane Loudon herself termed the Linnaean system "unfit for females" and contrasts it with the natural arrangement of plants, which has "nothing objectionable."[50]

Botany for Ladies (1842), Jane Loudon's botanical magnum opus, is a technical but accessible "Popular Introduction to the Natural System of Plants, according to the Classification of de Candolle." Written in two parts, it begins with descriptions of a few orders and genera that contain plants commonly grown in England and then presents a synopsis of Augustin-Pyramus de Candolle's system. Jane Loudon explains that the purpose of the book is to aid plant identification: "The grand object I have in view is to enable my readers to find out the name of a plant when they see it for the first time; or, if they hear or read the name of a plant to make that name intelligible to them."

Botany for Ladies is Loudon's alternative to John Lindley's *Ladies' Botany*, which had appeared eight years earlier. Lindley, professor of botany at London University, had included a botany book for women among the textbooks he wrote to introduce the natural system of plant classification to a new generation of readers; he wanted, as we have seen, to make his version of scientific botany accessible for women. But the level of botanical discourse in *Ladies' Botany* was still a problem for some female students. As Loudon put it, "even [*Ladies' Botany*] did not tell half I wanted to know, though it contained a great deal I could not understand." She eloquently pointed out the dilemma of a threshold for scientific exposition that was still too high for many women, herself included: "It is so difficult for men whose knowledge has grown with their growth, and strengthened with their strength, to imagine the state of profound ignorance in which a beginner is, that even their elementary books are like the old Eton Grammar when it was written in Latin—they require a master to explain them." [51] *Botany for Ladies* was written, by contrast, for women like herself.

In her own exposition of a botanical system, Jane Loudon, unlike the previous generation of female botanical writers, did not turn to the familiar format. She framed her discussion in the first person ("My readers may perhaps . . . be as much surprised as I was to learn that . . .") but included no sisterly conversations in a domestic setting; nor did she elaborate on literary or moral associations of plants or their culinary or medical uses. *Botany for Ladies* differs markedly in this regard from the conversational and epistolary, domestic narratives of women writers one generation earlier. At the beginning of her career she had written one book in the familiar format, *Conversations upon Chronology and General History, from the Creation of the World to the Birth of Christ* (1830), modeled on Jane Marcet's *Conversations on Natural Philosophy* and featuring a mother and two daughters. But none of Loudon's subsequent expository books introduces a maternal teacher or narrative mother. *First Book of Botany* (1841), an introductory book "for Schools and Young Persons," would have been an obvious venue for a family-based narrative in the conversational or epistolary mode, but her account of the basic terms of botany resembles instead a standardized textbook, with no cast of characters and no designated interlocutors.

By the 1840s the conversational or epistolary mode in women's scientific writing had little currency. Jane Loudon, seeking a scientific botany for beginners, sought an appropriate narrative form that would invite women readers into botanical study. She made the decision against the conversation and in favor of the title *Botany for Ladies*. She thereby continued to shelter under the safety of an older and familiar style of title while also representing a new format for women's botany writing. By writing an introductory book without mothers, she turned her back on a format that had dominated women's popular science writing since the late eighteenth century.

Loudon's choice of textual style involved more than a formal literary decision. It also reflected her own pedagogical ideas about girls' education. During the 1790s many writers advocated home education for girls instead of fashionable boarding schools. Writers as politically disparate as Priscilla Wakefield and Hannah More shared the conviction that mothers should take on a new centrality in educating their daughters. Jane Loudon took a different view in the 1840s, supporting home education for most girls but arguing against maternal tutelage in all areas beyond early moral training. She favored the professionalization of teaching, which meant bringing in specialist masters rather than relying on generalist mothers. As she put it in an essay for the *Ladies' Companion at Home and Abroad*, a large-format Saturday weekly she founded and edited, "I do not think that mothers can teach their children so well as those who have made teaching a study." [52] Loudon thereby promoted a teaching format that eliminated botany from mothers' pedagogical repertory. As specialist teachers took over, narrative mothers would become a detail in the literary history of early textbooks.

Loudon was a professional woman writer impelled by economic circumstances to find and keep audiences for her books. We can expect her, therefore, to have been responsive to changing tastes among readers and publishers. One recalls John S. Forsyth's attack in 1827 on the "garrulous old women [and] pedantic spinsters" who appeared in popular science books from the 1790s through the 1820s. By the 1840s narrative styles in science writing had changed, and the familiar format had become distinctly old-fashioned. Hence Jane Loudon's choice of a narrative form for her botany books was also partly linked to market forces. The death of the seemingly indefatigable John Claudius Loudon in 1843 gave his widow, then thirty-six years old and the mother of a young daughter, a powerful economic spur to her writing career. Although Loudon held many copyrights on his books, he left substantial publishing debts, and various agencies came to Jane Loudon's assistance. As "Widow of the late John Claudius Loudon, author of several works connected with botanical science," she was awarded an annual pension of one hundred pounds sterling from the Civil List for his "services and merits." Public meetings took place at the Horticultural Society to establish an annuity for the family by promoting the sale of John Claudius Loudon's work. The Royal Literary Fund invited her to apply and assisted her then, as they had earlier in her writing career.[53] In May 1844 she wrote: "I have at present no income but what arises from my literary activities but I have the hope of . . . income from the copyright of my late husband." She hoped "soon to be able to maintain the family by my literary exertions."[54] Over the next fifteen years she labored to support herself and her daughter Agnes by her writing, and she died relatively young.

Jane Loudon did not place herself dramatically at variance with mainstream cultural practices in either her botanical or her general writing. The natural system of plant classification had sufficiently replaced Linnaean botany by the 1840s that plant sexuality was not a pressing issue for introductory texts, nor did an author need to participate in controversies about the standing of competing taxonomic systems. But she wrote with an ideal paying readership always before her eyes and at her elbow and was cognizant of public taste (or at least the taste of those purchasing her books). Accordingly, she avoided reference to plant sexuality, choosing the bowdlerizing pathway laid down by Mrs. Perkins in her *Elements of Botany* from the 1830s. Her *First Book of Botany*, explaining branches of botany for "schools and young persons," made no reference at all to sexual reproduction.

Jane Loudon, a professional writer, insisted on female mental improvement, but within prevailing gender ideology about separate spheres for women and men. She filled her periodical the *Ladies' Companion* with articles about dress and fashion, book reviews for the family circle, patterns for tatting, biographical essays on women, and "household hints and receipts" and explained in the editor's statement in the first issue that she aimed "not to make women usurp the place of men, but to render them rational and intelligent beings." She considered religion and politics unsuitable subjects for her magazine. She also recalled having "a most awful idea of a learned lady, or 'blue stocking' " as a child during the opening decades of the nineteenth century; "I always pictured to myself . . . a cross old maid, who did not like children, and who talked in high-flown language that very few people could understand." [55]

Science, however, was fully appropriate, and the *Ladies' Companion* includes many articles on science, among them essays by male scientists on geology, physical geography, and chemistry and a series of fifteen essays on botany by Jane Loudon herself. In fact, Loudon's magazine for women shows her to have been a modernizer and professionalizer within the domains of knowledge in her day. Revealingly, for a writer steeped in the culture of plants and flowers, she showed no interest in the herbal tradition. Unlike her contemporary Anne Pratt, she included no discussions about the culinary or medical uses of plants and pointedly refused to give medical advice in her magazine. Replying to a reader in 1850, she wrote: "When I am ill, I always apply to a doctor, and I advise my correspondents to do the same." [56] There can be no greater evidence for the demise of the "wisewoman" among the urban middle class than that statement, in which a woman deeply versed in horticultural and botanical information distances herself from the traditional domain of women and herbal knowledge.

A deliberate modernizing direction in Loudon's writing became even more evident when a second edition of her *Botany for Ladies* appeared in 1851 under the title *Modern Botany*. In the year of the Great Exhibition at the Crystal Palace, her popular introductory botany book on the natural system of plant classification and identification received a new title for a new decade. The adjective "modern" no doubt increased the attractiveness of the book for authors, publishers, and book buyers who wanted to be up to date. Whether the decision about a new title originated with Jane Loudon, widow and author, or with her publisher, the change resonates with midcentury beliefs about reading audiences, education, the place of science, modernity, and the literary marketplace.

The decision to market Jane Loudon's botany book under the new title *Modern Botany* also reflects midcentury debates about women's education in relation to gender ideologies. Although widely held Victorian social ideals called for training women for domestic roles, motherhood, and conjugal responsibilities, the social reality of "redundant women" who did not marry aroused social concern, and a generation of schoolmistresses began to shape an education for middle-class girls that would prepare them for paid employment. In contrast to traditionalist educators who believed women should learn topics different from those taught to men,

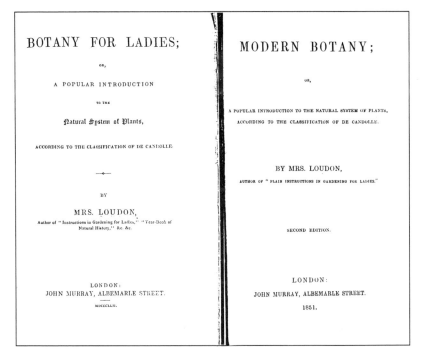

Jane Loudon, *Botany for Ladies* (1842); retitled *Modern Botany* in the edition of 1851.

on the assumption that differences in social roles dictated different methods of education, many women educators believed sex-differentiated models of education were not advantageous for girls, particularly for preparing middle-class girls for work in the world. They wanted instead to adopt the classical and gentlemanly ideal of liberal education that had been the model in boys' schools. They supported moral and intellectual training as a model for girls' education and incorporated the modern languages and science. By 1850 the practices of boys' schools were becoming the model for female education too.[57] Ideas like these informed Jane Loudon's writing as well as her approach to periodical publishing. An essay "Women's Books and Men's Books" in her magazine the *Ladies' Companion* rejects the view that women's reading and study should be confined to sex-specific topics and argues that women should read men's texts just as men should read magazines like the *Ladies' Companion*. The essayist's position was that, aside from different male-female duties and employments, there are topics that "have no sex."[58]

Within the context of discussions in her day about sex differences, women's education, and curricular reform, the change in title from *Botany for Ladies* into the non-sex-specific *Modern Botany* signaled interest in guiding women readers beyond a narrow notion of separate spheres into a wider orbit of mid-nineteenth-century students of botanical science. If "the mind has no sex," women's science education should not differ from men's, and women and men should read the same science books. The freshly titled *Modern Botany* thus can be seen as opening a broader intellectual space for women while also inviting male readers into this introductory text.

Yet the disappearance of books designated specifically for them erases women from the public stage of science—at least in books. Women's presence fades from the surface of botanical texts, and women's spaces also disappear as the site of science. These changes reflected dramatic transformations in the self-representation of botany, away from home- and family-centered botany as leisure activity and away from nature study as part of general education. They mark an increasing turn toward narratives of science.

A Feminist's Flora: Lydia Becker

In 1864 Lydia Ernestine Becker (1827–90), soon to become a prominent feminist and suffrage leader, published *Botany for Novices: A Short Outline of the Natural System of Classification of Plants.* This book teaches principles of botany by reference not to names but to structure. It focuses on flowering plants within the natural system and provides a clear and simple exposition of differences between dicotyledons and monocotyledons within that system. The author explains how to dissect and observe parts of plants; comparative remarks about growth rings in a woody horse chestnut and a cane plant, for example, lead into discussion of exogenous and endogenous structures. This "little book" is directed to "those persons, who, while wishing to know something of Botany, feel deterred by the long words and apparently formidable difficulties which meet their perplexed gaze on opening an elaborate treatise on the subject; or who entertain the groundless fear that they cannot learn much of the science without burdening the memory with a great many long, hard names of plants, which will leave them, after the trouble of learning them, little wiser than before."[59]

Botany for Novices, by an author identified only as "L.E.B.," appeared in the same year as Daniel Oliver's "modern" version of J. S. Henslow's botany book and, like it, contains a representation of botanical science.

Portrait of Lydia Becker by Susan Isabel Dacre. (Courtesy of Manchester City Art Gallery)

The book was intended especially for young women, but nothing in either the preface or the narrative of *Botany for Novices* makes that explicit. Addressed to a non-sex-specific audience, the exposition is exclusively botanical, with no moral teachings, spiritual references, or miscellaneous educational comments, nor are there interpolated poems. Although written by a woman, the book is no different from a standardized, impersonal male botany book. Indeed the author, Lydia Becker, would not have wanted her book to be identified as by a woman or exclusively designated for a woman-centered audience.

Lydia Becker was a well-known Victorian feminist, paid secretary of the National Society for Women's Suffrage from 1867 and editor of its monthly magazine, the *Women's Suffrage Journal* (1870–90).[60] An energetic lecturer and essayist who addressed the status of women in all its aspects, she did detailed daily work for the suffrage campaign over many years. In 1868 she helped form the Married Women's Property Committee, which worked to pressure Parliament toward legal reform in that area. She was elected to the new Manchester school board and served as part of the first wave of women in local public office, taking a special interest in the education of girls.[61] She was president of the Manchester Ladies' Literary Society and organized a lecture series for women.

Becker was an eloquent advocate of science education for women during decades when schools were replacing home-based education for the middle classes and when opportunities for female science education were opening up in some schools and universities.[62] Science education for girls was put on the national educational agenda in England in 1868 through the work of the Schools' Enquiry Commission, which included girls' schools in its terms of reference. The report of what became known as the Taunton Commission criticized girls' education as inadequate. Curricular and pedagogical discussions ensued about elementary and higher education for girls. Lydia Becker fervently supported and promoted the view that education for women should be based on a unified standard rather than on notions of difference and separate spheres. She saw no intrinsic differences between the female mind and the male mind. Hence she argued that single-sex education at university level and "supernumerary 'women's classes'" were "a departure from the principle of equality." She opposed sex segregation in teaching women science at the university level and wrote: "In speaking of the study of science by women, I desire, at the outset, to guard against the supposition that I consider such study to present any exceptional peculiarity to distinguish it from the study of science by men. Male and female students, in any branch of science, must go through the same training, and have their qualifications and capacities tested by precisely the same rules; neither is there anything in these studies which is naturally more attractive or advantageous to persons of one sex than of the other."[63]

Becker's style of mid-Victorian feminism supplies the theory and the strategy that underlay *Botany for Novices*. Earlier botany books directed to women assumed that female education in botany took place most successfully through a textual separate sphere. Gender-specific texts for women varied in their voices, from the maternal to the patronizing to the sisterly, but they were usually alike, as I have shown, in choosing formats associated with the private world of home or personal relations. Becker would spurn the familiar format in introductory botany books as too exclusively female identified. Probably more important, she also thought it was strategically correct to teach women science as it was being taught to men. In her view, success lay in putting the female—and certainly the feminine—aside, disallowing sex differences in education and in pedagogy.

Lydia Becker grew up in the Lancashire countryside as the eldest child of a large and prosperous manufacturing family and was educated at home. She became active in botanical work as a young woman, encouraged by an uncle who was enthusiastic about botany. John Leigh, an important figure in Manchester scientific culture, corresponded with her about plants, sent her specimens, tutored her on how to collect, dry, and exchange plants, and gave advice about botany books for her to read. Leigh seems to have set Becker on a serious botanical course, particularly toward studying plant structure rather than just classifying her specimens.[64] During the 1860s she won a national prize for a collection of dried plants (having devised a system for drying that retained colors well). She corresponded during the years 1863–77 with Charles Darwin about botanical topics such as variation and hybridity and sent him plants. (For the first meeting of the Manchester Ladies' Literary Society she invited Darwin to send a paper about botany, and he obliged by sending two.)[65] She herself conducted botanical research on the effect of a fungus parasite on the sexual development of a particular plant. In a botanical paper presented at the 1869 Exeter meeting of the British Association for the Advancement of Science, she reported that "the parasitic fungus [of *Lychnis dioica*] caused the flowers to assume . . . bisexual form"; she suggested that this might illustrate Darwin's theory of pangenesis.[66]

Botany remained an important part of her private life even as she immersed herself ever more deeply in work on behalf of suffrage. When Charles Babington, professor of botany at Cambridge, declined signing a suffrage petition she sent him in 1867, he wrote, "I hope that after this matter is settled you will return to the gentle science of Botany and that I may have the opportunity of helping you in it."[67] Lydia Becker's ardor for botany remained strong. "I hope I do not tire you with [botanical] details," she wrote in 1887, "but plants are my hobby, and when I am mounted I never know when to stop."[68] Her biographer reported that her "best refreshment" when political work became too hectic was "a run down to the gardens and conservatories at Kew."[69]

Botany for Novices embodies the cultural practice that Lydia Becker worked for. Becker, a feminist botanist, accepted masculine science as universal science. Her strategy was probably the right one for women at that time, since sex-differentiated practices in her day translated into a gendered hierarchy and a sexual politics disadvantageous to women. So long as sexual difference was construed as female inferiority, a rhetoric of educational equality was an important strategy. Becker based her argument for male-female equality and for science education on assumptions about sameness and thereby sought to ensure that at least some women had access to university science. This strategy would open the pathways of professionalism to some. Like John Lindley several decades before, Lydia Becker opposed a model of politeness and promoted women's study of botany on the more serious male model. Opposing sex segregation, however, meant accepting the sex-gender-science male model of science and of the "scientist."

Compared with Priscilla Wakefield, Becker gave female readers of *Botany for Novices* a much wider field for knowledge, well beyond home and mother, with no geographical or mental restrictions. But her science is depersonalized and decontextualized: science and life have been separated, and there are no longer any textual links between science and morals. The early botany books by women showed a world of subjects in which science was part of a humanistic enterprise. Priscilla Wakefield, Sarah Fitton, Jane Marcet, even Jane Loudon in the 1840s represent personalized voices. The introductory books that came to be the model in botanical education, and which Lydia Becker accepted for her own feminist reasons, replaced all this with a representation of science that lacked a visible shaper and with a disembodied voice of (ostensibly) invisible and inaudible gender.

The issue remains a problem. In a professionalizing and defeminizing science culture, it may seem advantageous to adopt the equality model and ensure that women have access to mainstream science. But that is science constructed on the male model, far removed from family, home, mothers, contextual thinking, and other cultural features that have become gendered as "feminine."

Flora Feministica

On December 3, 1896, Beatrix Potter, a pioneering researcher in mycology and later the author and illustrator of *The Tale of Peter Rabbit*, sat outside the office of William T. Thiselton-Dyer, director of the Royal Botanic Gardens, Kew, waiting to be admitted. She had succeeded in sprouting the spores of one kind of fungi and was carrying a report about spore germination. Six months earlier, accompanied by her uncle, the distinguished chemist Sir Henry Roscoe, she had shown her fungus drawings to the director, who "seemed pleased, and a little surprised." Now, at Kew by herself, she saw Thiselton-Dyer through a window, but after a while she was "so seized with shyness" that she "bolted." She recorded all this in code in her diary and added that there was "half-thawed ice on the pond and terrible mud." Another visit followed in which Thiselton-Dyer dismissed her in a rude and condescending way; she wrote, "I fancy he may be something of a misogynist."[1]

Beatrix Potter made another attempt to gain a hearing for her research about spore development in fungi when her paper "On the Germination of the Spores of *Agaricineae*" (by Miss Helen B. Potter) was "communicated" (possibly only by its title) by proxy at a meeting of the Linnean Society, London, on April 1, 1897. Since women were not allowed to attend its meetings and present their own research, she could only hear at second hand that her paper was well received. Publication normally followed upon a presentation to the Linnean Society, but Beatrix Potter's paper never appeared. She reportedly was told that it needed more work, but the circumstances remain unclear. She continued to make detailed watercolors of fungi using high degrees of magnification but made no further forays into the professionalizing culture of establishment botany. Peter Rabbit and the hedgehog Mrs. Tiggy-Winkle became her conversationalists of choice, and her skills as a scientific illustrator fed into her classics of children's literature, with their tales of animal life and renderings of fungi, ferns, and landscapes.[2]

Beatrix Potter, age twenty-six, photographed by A. F. Mackenzie. (Photograph courtesy of the Victoria & Albert Museum, c/o Frederick Warne & Co. Reproduced by permission of Frederick Warne & Co.)

Beatrix Potter (1866–1943) engaged avidly in botanical research yet, like Agnes Ibbetson during the opening decades of the nineteenth century, was frustrated in efforts to have her contributions acknowledged by institutionally based botanists. She came into botany in the way many earlier botanical women had, as part of polite culture and within routines of mid-Victorian bourgeois life. Educated at home in London, with access to drawing teachers, she is said never to have gone anywhere alone except to the nearby Natural History Museum. She collected fungi during family holidays in Scotland and the Lake District, drew plants, and dissected, identified, and classified them. Her pleasure and absorption in fungi recall Anna Maria Hussey's mycological writing and illustrations fifty years earlier. Isolation provided many opportunities for natural history study, and Beatrix Potter's interest in scientific study grew well beyond rational recreation. Her scientist-uncle, who was also vice-chancellor of London University, served as mentor and patron.[3]

During the early days of polite botanical culture, women had a relatively smooth path into botanical conversations as interlocutors in botanical dialogues or as teachers, writers, and audience for public lectures. From Flora and Ingeana in the *New Lady's Magazine* of 1786 to Hortensia and her children in Maria Jacson's *Botanical Dialogues* of 1797, from the duchess of Portland and Mary Delany enthusing about puffballs in the drawing room to Sarah Abbot talking with her clergyman-husband about plants for his flora of Bedfordshire, to the few women members of the Botanical Society of London and Elizabeth Warren's epistolary exchanges with William Hooker about her botanical chart, women had botanical conversations in print and in person and claimed botany as a resource for themselves. This was the case especially from the 1780s until that time, roughly sixty years later, when academic botany set itself at a remove from popular pursuits and specialized science endeavored to supplant polite science culture. During the Enlightenment, the linkage between scientific activities and polite activities had enabled women to elbow into science culture. Later that same linkage led influential practitioners to elbow, or edge, women out, as cultural connections between botany and femininity provoked anxieties among some male botanists and some male popular science writers concerned to construct a botanical science. The professionalizing turn within science culture, indeed within nineteenth-century culture more generally, marginalized many activities and values associated with women. After the 1860s, the locus of learning shifted from home-based and mother-centered education to laboratory-based school science. Pedagogical authority changed from a mentorial mother to a formal schoolteacher in a public institution, and examination systems institutionalized a certain kind of knowledge, marginalizing the autodidact.[4]

The parting of the ways between professionals and amateurs, and between science based in the laboratory and science based in the field or at home, was not exclusively disadvantageous to women, however. Developments in late nineteenth-century culture gave some women access to scientific subjects at schools, universities, and medical schools; over six hundred women received B.Sc. degrees at the University of London between 1879 and 1911, for example, and women wrote the natural science tripos at Cambridge, some going thereafter into careers as schoolmistresses.[5] Read within this context, the rebuff to Beatrix Potter at Kew and the Linnean Society in 1896 and 1897 shows her to have been a late casualty to older ways in institutional botany, for serious lobbying began in 1900 toward admitting women to membership in the Linnean Society; the portals finally opened in 1904, when sixteen women were proposed for election as fellows.[6]

The next generation of women in botanical culture included academic teachers and researchers as well as writers who bridged the gap between specialists and general readers. Marie Carmichael Stopes (1880–1958), for example, a paleobotanist (and pioneering sex educator and birth-control advocate), lectured at University College, London, where John Lindley had been the first professor of botany and where she herself received a B.Sc. with double honors in botany and geology. After doctoral work in Germany, she was appointed junior lecturer and demonstrator in botany at Manchester University, the first woman on the science staff. She wrote many technical papers and monographs on fossil plants, including cycads, and linked paleobotany and coal research. She also wrote elementary science books. *The Study of Plant Life for Young People* (1906), her first book, and *Ancient Plants: Being a Simple Account of the Past Vegetation of the Earth* (1910), written for the nonspecialist, locate Marie Stopes in the tradition of women working as cultural mediators in science.[7] In particular Marie Stopes, educator, belongs to the history of women who, since the later eighteenth century, wrote for a juvenile audience and created styles of exposition for readers on the threshold of science. In her day there was also a new generation of women popularizers who, constructing narrative authority in new ways, wrote books about science that spoke to experts about their fields of specialization. Other women wrote books of general interest, books for children, and formal school textbooks. Within botanical culture, Charlotte L. Laurie, assistant mistress at Cheltenham Ladies' College, wrote *Flowering Plants: Their Structure and Habitat* (1903), for example; and Eleanor Hughes-Gibbs wrote *The Making of a Daisy, "Wheat out of Lilies," and Other Studies in Plant-Life and Evolution: A Popular Introduction to Botany* (1898), advocating "friendship with Mother Nature."

Botanists now immersed in technical books and professional abstracts about electron microscopy, chemotaxonomy, and molecular biological study may find little to recognize in the text world of earlier botanical culture. The essays of Agnes Ibbetson probably read to them like pages from the musty album of preprofessional science, with no bearing on current scientific practices. They likely would assign books by Maria Jacson, Anne Pratt, and Jane Loudon only to a cultural pile marked "literary" rather than "scientific." And books in the familiar format of letters and conversations probably appear very old-fashioned indeed.

Yet science writing has a history, and one feature is the disappearance of the narrator from science books, the erasure of personality from representations of science. The emphasis on maternal science education in the late eighteenth century gave latitude for women to study science and develop expertise. Some women harnessed social conventions and made scientific motherhood a career by writing books in the textbook tradition. They shaped a women-centered science pedagogy that provided a form of intellectual authority. Narrative mothers in the female mentorial tradition taught science and represented female scientific literacy. The textual exclusion of women and the family from the new-style formalized science book by about 1850 marginalized one area of women's experience. Although some women in later years had enhanced access to mainstream science, it probably was the case that more women took part in scientific work when it was in the domestic sphere than when it moved away from home.

This book about botanical conversations is part of our continuing conversation with the historical past. As educators and policy makers work to increase access to science for girls and women, we would do well to examine not only contemporary gender issues but also the gendered history of scientific disciplines and the pervasive factor of gender in shaping the scientist, science education, and science writing. Botany, for one, should acknowledge its female and feminine parts. A broader conversation about the culture of botany would include more speakers and more topics of conversation, welcoming diversity in backgrounds and agendas. During the Enlightenment, scientific education was linked to beliefs about what Priscilla Wakefield and others called "mental improvement," and science and ethics were joined; writers did not distinguish sharply between science and art, specialist knowledge and generalist knowledge, technical writing and popular writing. In our day, when the culture clash between the sciences and humanities is acute, conversations between Flora and Ingeana, and our conversations about nature and science, women and gender, are more vital than ever.

Notes

Prologue: Botanical Conversations

1. *New Lady's Magazine, or Polite and Entertaining Companion for the Fair Sex*, May 1786, 177.

2. See Gerald Dennis Meyer, *The Scientific Lady in England, 1650–1760: An Account of Her Rise, with Emphasis on the Major Roles of the Telescope and Microscope* (Berkeley and Los Angeles: U of California P, 1955); John Mullen, "Gendered Knowledge, Gendered Minds: Women and Newtonianism, 1690–1760," in *A Question of Identity: Women, Science, and Literature*, ed. Marina Benjamin (New Brunswick: Rutgers UP, 1993); Patricia Phillips, *The Scientific Lady: A Social History of Woman's Scientific Interests, 1520–1918* (London: Weidenfeld and Nicolson, 1990), chap. 5; G. S. Rousseau, "Scientific Books and Their Readers in the Eighteenth Century," in *Books and Their Readers*, ed. Isobel Rivers (New York: St. Martin's P, 1982), 212–14; Teri Perl, "The Ladies' Diary or Woman's Almanack, 1704–1841," *Historia Mathematica* 6 (1979): 36–53; and Aileen Douglas, "Popular Science and the Representation of Women: Fontenelle and After," *Eighteenth Century Life* 18 (May 1994): 1–14.

3. For an ethnographic and historical approach to this subject, see Jack Goody, *The Culture of Flowers* (Cambridge: Cambridge UP, 1993).

4. John Lindley, *An Introductory Lecture Delivered in the University of London on Thursday, April 30, 1829* (London, 1829), 17.

5. See, e.g., F. W. Oliver, ed., *Makers of British Botany: A Collection of Biographies by Living Botanists* (Cambridge: Cambridge UP, 1913); Ellison Hawks, *Pioneers of Plant Study* (London: Sheldon, 1928); and A. G. Morton, *History of Botanical Science: An Account of the Development of Botany from Ancient Times to the Present Day* (London: Academic, 1981).

6. On the social history of natural history in England, see David Elliston Allen, *The Naturalist in Britain: A Social History* (1976; 2d ed., Princeton: Princeton UP, 1994), and numerous essays; also Nicolette Scourse, *Victorians and Their Flowers* (London: Croom Helm, 1983). On women and botany in Sweden, see Gunnar Broberg, "Fruntimmersbotaniken" ["Botany for women"], *Svenska Linnésällskapets Årsskrift* 12 (1990–91): 177–231.

7. See Pnina Abir-Am and Dorinda Outram, eds., *Uneasy Careers and Intimate Lives: Women in Science, 1789–1979* (New Brunswick: Rutgers UP, 1987); Phillips, *Scientific Lady*; Elizabeth B. Keeney, *The Botanizers: Amateur Scientists in Nineteenth-Century America* (Chapel Hill: U of North Carolina P, 1992). Also see Londa Schie-

binger, *The Mind Has No Sex? Women in the Origins of Modern Science* (Cambridge: Harvard UP, 1989).

8. See David Locke, *Science as Writing* (New Haven: Yale UP, 1992); Greg Myers, "Science for Women and Children: The Dialogue of Popular Science in the Nineteenth Century," in *Nature Transfigured: Science and Literature, 1700–1900*, ed. John Christie and Sally Shuttleworth (Manchester: Manchester UP, 1989), and idem, "Fictions for Facts: The Form and Authority of the Scientific Dialogue," *History of Science* 30, 3 (1992): 221–47; Charles Bazerman, *Shaping Written Knowledge: The Genre and Activity of the Experimental Article in Science* (Madison: U of Wisconsin P, 1988); and Steven Shapin and Simon Schaffer, *Leviathan and the Air-Pump: Hobbes, Boyle and the Experimental Life* (Princeton: Princeton UP, 1985), esp. chap. 2.

9. See Margaret J. M. Ezell, *Writing Women's Literary History* (Baltimore: Johns Hopkins UP, 1993).

10. Marina Benjamin, "Elbow Room: Women Writers on Science, 1790–1840" in *Science and Sensibility: Gender and Scientific Enquiry, 1780–1945* (Oxford: Blackwell, 1991); Barbara T. Gates, "Retelling the Story of Science," in *Victorian Literature and Culture*, vol. 21, ed. John Maynard and Adrienne Auslander Munich (New York: AMS Press, 1993), 289–306; and Barbara T. Gates and Ann B. Shteir, *Science in the Vernacular*, forthcoming.

Chapter 1. Spreading Botanical Knowledge

1. James Sowerby, *Botanical Pastimes . . . Calculated to Facilitate the Study of the Elements of Botany* (London: Darton, Harvey, and Darton, ca. 1810). Card number one in the pack, for example, asks for the "description of a perfect Plant." The player holding the answer card ("It consists of the root, the trunk or stem, the leaves, supports, flower, and fruit") in turn asks a question. The player who answers incorrectly receives a card from the interrogator. The winner is the first player to dispose of all cards (Natural History Museum, London).

2. On eighteenth-century science in England see Larry Stewart, *The Rise of Public Science: Rhetoric, Technology, and Natural Philosophy in Newtonian Britain, 1660–1760* (New York: Cambridge UP, 1992); James E. McClellan, *Science Reorganized: Scientific Societies in the Eighteenth Century* (New York: Columbia UP, 1985); and Maureen McNeil, *Under the Banner of Science: Erasmus Darwin and His Age* (Manchester: Manchester UP, 1987).

3. For a schematic account of the Linnaean system, see William T. Stearn's appendix to Wilfrid Blunt, *The Compleat Naturalist: A Life of Linnaeus* (London: Collins, 1971), 242–49. See also Gunnar Eriksson, "Linnaeus the Botanist," in *Linnaeus: The Man and His Work*, ed. Tore Frängsmyr (Berkeley and Los Angeles: U of California P, 1983).

4. This poem, in manuscript on the endpapers of a British Library copy of the (1783) translation by the Botanical Society of Lichfield of Linnaeus's *A System of Vegetables*, vol. 2 (BL447c.19), probably was written as an occasional poem to mark the appearance of the translation. Although the poet is unidentified, internal evidence suggests Anna Seward.

5. On the early history of sexual reproduction in plants, see John Farley, *Gametes*

and Spores: Ideas about Sexual Reproduction, 1750–1914 (Baltimore: Johns Hopkins UP, 1982).

6. Linnaeus, *A Dissertation on the Sexes of Plants*, trans. James Edward Smith (London, 1786).

7. See Londa Schiebinger, *Nature's Body: Gender in the Making of Modern Science* (Boston: Beacon, 1993), chap. 1.

8. See Thomas Laqueur's theory about the rise of the two-sex system in science in *Making Sex: Body and Gender from the Greeks to Freud* (Cambridge: Harvard UP, 1990), chap. 5; and Londa Schiebinger's theory about the development of the theory of complementarity in *The Mind Has No Sex?* chaps. 7 and 8.

9. On gender anxieties in relation to "Macaroni manners" during the 1770s, see Paul Langford, *A Polite and Commercial People: England, 1727–1783* (Oxford: Oxford UP, 1992), chap. 12. See also G. J. Barker-Benfield, *The Culture of Sensibility: Sex and Society in Eighteenth-Century Britain* (Chicago: U of Chicago P, 1992).

10. Charles Alston, "A Dissertation on the Sexes of Plants," in *Essays and Observations, Physical and Literary*, 2d ed. (Edinburgh, 1771), 1:263–316.

11. Letter dated November 16, 1821, in Lady Pleasance Smith, ed., *Memoir and Correspondence of the Late Sir James Edward Smith, M.D.* (London, 1832), 1:507. See also Frans A. Stafleu, *Linnaeus and the Linnaeans: The Spreading of Their Ideas in Systematic Botany, 1735–1789* (Utrecht: International Association for Plant Taxonomy, 1971).

12. A newspaper advertisement for one such field trip read: "On Saturday the 31st instant, W. Curtis, Demonstrator of Botany to the Company of Apothecaries, and Author of the Flora Londinensis, will make his First Herborizing Excursion, for this season, to the fields and meadows about Battersea. To meet at the Nine Elms coffee-house, near Vauxhall, at one o'clock. Such Gentlemen as wish to acquire a knowledge of the Plants growing wild about town, or of Linnaeus's System of Botany, have now a favourable opportunity, and may become pupils on Saturday next" (*Gazetteer*, May 28, 1777). Curtis established the London Botanic Garden as a private garden at Lambeth in 1779 and moved it ten years later to Brompton; in 1801 it had 213 subscribers. See Ray Desmond, *A Celebration of Flowers: Two Hundred Years of Curtis's Botanical Magazine* (Kew: Royal Botanic Gardens, 1987), 21.

13. Desmond, *Celebration of Flowers*, 40, 23.

14. The recipient of the letters, Madeleine-Catherine Delessert, became rather more of a botanist than her mentor would have welcomed; Blanche Henrey described her as "a keen botanist and the owner of a famous herbarium and botanical library." *British Botanical and Horticultural Literature before 1800* (London: Oxford UP, 1975), 2:55.

15. Jean-Jacques Rousseau, *Letters on the Elements of Botany*, trans. Thomas Martyn, 2d ed. (London, 1787), v, 18, 86.

16. Allen, *Naturalist in Britain*, 48. The second edition appeared 1787–92. The third edition carried a new title: *An Arrangement of British Plants; According to the Latest Improvements of the Linnean System* (1796). After Withering died, his son oversaw publication of three further editions of the book, which continued to appear until the fourteenth edition in 1877.

17. Mary Moorman ed., *Journals of Dorothy Wordsworth* (London: Oxford UP,

1971), 16, n. 2. The extant four-volume edition carries their marginal notes. See John Blackwood, "The Wordsworths' Book of Botany," *Country Life*, October 27, 1983, 1172–73; and D. E. Coombe, "The Wordsworths and Botany," *Notes and Queries* 197 (1952): 298–99.

18. In Sarah Wilson's *A Visit to Grove Cottage* (1823) a London-bred young woman visiting the daughters of country friends complains that "botany appears to me to be such a very dry, unentertaining study, full of such hard names and difficult words, that, much as I love plants and flowers, I do not think I should ever like it; besides, I do not know of what *use* all those long words are, such as *Monandria* and *Diandria*, which I have sometimes seen in your *Withering.*" Emmeline, the country daughter who loves botany, rises to Withering's defense (34–35).

19. See Robert E. Schofield, *The Lunar Society of Birmingham: A Social History of Provincial Science and Industry in Eighteenth-Century England* (Oxford: Clarendon, 1963).

20. William Withering, *Miscellaneous Tracts: To Which Is Prefixed a Memoir of His Life, Character, and Writings*, ed. William Withering the Younger, 2 vols. (London, 1822), 1:113.

21. For a window on the issues see Desmond King-Hele, ed., *The Letters of Erasmus Darwin* (Cambridge: Cambridge UP, 1981), 111–15. During the early 1780s the two men engaged in a series of botanical and personal disputes. Withering was well known for his medical treatises on scarlet fever and digitalis; his *Account of the Foxglove and Some of Its Medical Uses* (1785) reported on ten years of clinical observations. Darwin claimed that his own son, recently deceased, had made the first discovery about digitalis, and Withering was offended by this claim. See Schofield, *Lunar Society*, 163–65, and Desmond King-Hele, *Doctor of Revolution: The Life and Genius of Erasmus Darwin* (London: Faber and Faber, 1977), 101–3 passim.

22. This work is conventionally assigned to Erasmus Darwin as the chief member of an informal group whose other members were John Jackson and Brooke Boothby.

23. Anna Seward, *Memoirs of Dr. Darwin* (London, 1804), 167–68.

24. Erasmus Darwin, *The Botanic Garden* (1789–91; rpt. Menston, Yorkshire: Scolar, 1973), 9–10.

25. As Janet Browne showed, botany carried many different social and personal uses and meanings for Erasmus Darwin during the 1780s and 1790s. On the level of scientific politics, "The Loves of the Plants" was a weapon in the struggle between Linnaeans and anti-Linnaeans, by demonstrating the "natural" sexuality of plants and human beings. The emphasis on fecundity also expresses Erasmus Darwin's beliefs in transformation and progressive evolution in the natural world. On a personal level, Darwin may have intended the poem as "a kind of love song" to Elizabeth Pole, whom he later married. "Botany for Gentlemen: Erasmus Darwin and the Loves of the Plants," *Isis* 80, 4 (1989): 593–620. See also Londa Schiebinger, "The Private Life of Plants: Sexual Politics in Carl Linnaeus and Erasmus Darwin," in Benjamin, *Science and Sensibility*.

26. In *Plan for the Conduct of Female Education* Darwin's suggested bibliography for botany includes James Lee's *Introduction to Botany* and "the translations of the

works of Linnaeus by a society at Lichfield" (that is, his own). He particularly suggests Maria Jacson's *Botanical Dialogues between Hortensia and Her Four Children*, "a new treatise introductory to botany . . . for the use of schools, well adapted to this purpose, written by M. E. Jacson, a lady well skill'd in botany, and published by Johnson, London" (London: Johnson, 1797; rpt. New York: Johnson, 1968), 41.

27. Richard Polwhele, *The Unsex'd Females* (1798), 9, n. 1, 8–9. On Polwhele, see Emily L. de Montluzin, *The Anti-Jacobins, 1798–1800: The Early Contributors to the "Anti-Jacobin Review"* (New York: St. Martin's, 1988), 129–32.

28. Cf. Sandra M. Gilbert and Susan Gubar in *The Madwoman in the Attic* (New Haven: Yale UP, 1979), 7, exploring the metaphor of literary paternity and its implications for literary women, ask, "If the pen is a metaphorical penis, with what organ can females generate texts?"

29. Mary Wollstonecraft scathingly dismissed as "a gross idea of modesty" the reply of a journal editor who was asked by a female correspondent "whether women may be instructed in the modern system of botany consistently with female delicacy" and answered, "they cannot." *A Vindication of the Rights of Woman* (1792; London: Penguin, 1992), 293, 233.

Mary Wollstonecraft translated *Elements of Morality, for the Use of Children* (London: Johnson, 1791), a German collection of moral tales on such topics as slovenliness, honesty, and the value of work. The author, C. G. Salzmann, was interested in sex education: "I am thoroughly persuaded that the most efficacious method to root out [impurity] . . . would be to speak to children of the organs of generation as freely as we speak of the other parts of the body, and explain to them the noble use which they were designed for, and how they may be injured (xiv–xv).

30. Anna Seward, friend to Erasmus Darwin, is responding to criticisms of Darwin's posthumously published *Temple of Nature* as "not fit for the female eye." Letter 14 to Dr. Lister, from Lichfield, June 20, 1803, in Anna Seward, *Letters Written between the Years 1784 and 1807* (Edinburgh, 1811), 6:79–85.

31. Letter dated June 1, 1806; Linnean Society, Smith MSS, 19:176.

32. Mary Louise Pratt, *Imperial Eyes: Travel Writing and Transculturation* (London: Routledge, 1992), 25.

33. *The Autobiography and Correspondence of Mary Granville, Mrs. Delaney*, ed. Lady Llanover, ser. 1 (London, 1861), 1:559.

34. See Olivia Smith, *The Politics of Language, 1791–1819* (Oxford: Clarendon, 1984).

35. See Donald C. Goellnicht, *The Poet-Physician: Keats and Medical Science* (Pittsburgh: U of Pittsburgh P, 1984), and Hermione de Almeida, *Romantic Medicine and John Keats* (New York: Oxford UP, 1991). Also see Alan Bewell, "Keats's 'Realm of Flora,'" *Studies in Romanticism* 31, 1 (1992): 71–98.

36. In notes on a French botanist's work, Coleridge wrote: "what is BOTANY at this present hour? Little more than an enormous nomenclature; a huge catalogue, *bien arrangé.* . . . The terms system, method, science, are mere improprieties of courtesy, when applied to a mass enlarging by endless appositions, but without a nerve that oscillates, or a pulse that throbs, in sign of *growth* or inward sympathy."

Cited in Trevor Levere, *Poetry Realized in Nature: Samuel Taylor Coleridge and Early Nineteenth-Century Science* (Cambridge: Cambridge UP, 1981), 89.

37. *Annals of Philosophy* 16 (1820): 130.

Chapter 2. Women in the Polite Culture of Botany

1. *Gentleman's Magazine* 71, 1 (1801): 199–200.

2. Hannah More, *Strictures on the Modern System of Female Education* (1799), in *The Works of Hannah More* (London, 1853), 3:267.

3. See Richard Leppert, *Music and Image: Domesticity, Ideology and Socio-cultural Formation in Eighteenth-Century England* (Cambridge: Cambridge UP, 1988), esp. chap. 8.

4. *The Lady's Poetical Magazine, or Beauties of British Poetry* 1 (1781): 1–4. Queen Charlotte wrote in June 1779, "I am of [the] opinion that if women had the same advantages as men in their education they would do as well." She concerned herself with the education of her children and grandchildren, recommended the writings of Madame de Genlis, and received that well-known French educator and author at Windsor Castle. She wrote at one point that she was planning to introduce her two eldest daughters to some instruction in electricity and pneumatics, to help them get "quelques petites idées de Phisique." Olwen Hedley, *Queen Charlotte* (London: Murray, 1975), 125–26.

5. Smith, *Memoir and Correspondence*, 1:289–90.

6. Cited in Hedley, *Queen Charlotte*, 113, 181.

7. See Ruth Ginzburg, "Uncovering Gynocentric Science," in *Feminism and Science*, ed. Nancy Tuana (Bloomington: Indiana UP, 1989), 72.

8. See Agnes Arber, *Herbals, Their Origin and Evolution: A Chapter in the History of Botany, 1740–1670*, 3d ed. (Cambridge: Cambridge UP, 1986).

9. Lady Grace Mildmay exemplifies the many unlicensed women who were "unauthorized healers for their families and communities." Linda Pollock, *With Faith and Physic: The Life of a Tudor Gentlewoman, Lady Grace Mildmay, 1552–1620* (London: Collins and Brown, 1993).

10. Jane Barker, *A Patch-Work Screen for the Ladies* (1723; rpt. New York: Garland, 1973), 10, 56.

11. On women's medical cookery and healing practices, see Schiebinger, *The Mind Has No Sex?* 112–16; Elaine Hobby, *Virtue of Necessity: English Women's Writing, 1649–1688* (London: Virago, 1988), 177–78; Mary Fissell, *Patients, Power, and the Poor in Eighteenth-Century Bristol* (Cambridge: Cambridge UP, 1991), chaps. 2 and 3. Also Roy Porter, *Health for Sale: Quackery in England, 1650–1850* (Manchester: Manchester UP, 1989), and Dorothy Porter and Roy Porter, *Patient's Progress: Doctors and Doctoring in Eighteenth-Century England* (Oxford: Polity, 1989).

12. *The Works of George Farquhar*, ed. Shirley Strum Kenny (Oxford: Clarendon, 1988), vol. 2, 1:72–77; 4, 1:61–62.

13. Letter from "Philo-Naturae," *Female Spectator* (Glasgow) 4 (1775): 27–28.

14. Blanche Henrey recounts this detail in *British Botanical and Horticultural Lit-*

erature, 2:256. See also John Gascoigne, *Joseph Banks and the English Enlightenment: Useful Knowledge and Polite Culture* (Cambridge: Cambridge UP, 1994), 83.

15. Bridget Hill, *Women, Work, and Sexual Politics in Eighteenth-Century England* (Oxford: Blackwell, 1989), 163.

16. The philanthropic direction for women's herbal knowledge did not vanish. Anna Williams, poet and friend to Samuel Johnson, fantasized about her ideal house and activities in her poem "The Wish." Instead of "mean ambition" or "swelling ostentation," she wrote that one hundred pounds a year "Twill keep my friend, my maid, and me" in the kind of life she would like:

Sequestered from the curious eye
Should stand my little library,
Where free my mornings might remain
From interruption, kind or vain;
There good divines should Heaven impart,
And Poets humanize the heart;
Philosophy improve the mind,
And Physic teach to aid mankind.

.   .   .   .

My closet with choice med'cines stor'd,
Should to the poor relief afford.
    (*Miscellanies in Prose and Verse* [London, 1766])

Her dream of the calm and useful life, then, included a tradition of benevolent herbal practice.

17. Elizabeth Blackwell, *A Curious Herbal*, 2 vols. (London, 1737–39), text accompanying plates 379 and 497. Sir Joseph Banks's copy of *A Curious Herbal* (BL452 f.1) carries his annotations that identify Elizabeth Blackwell's plants according to the Linnaean system.

18. Blanche Henrey has reconstructed Elizabeth Blackwell's story in *British Botanical and Horticultural Literature*, 2:228–36. See also *Gentleman's Magazine* 17 (1747): 424–26.

19. See Wilfrid Blunt and Sandra Raphael, *The Illustrated Herbal* (London: Lincoln, 1979), 175–76.

20. See Gerta Calmann, *Ehret, Flower Painter Extraordinary: An Illustrated Biography* (Oxford: Phaidon, 1977).

21. "A Memoir of George Dionysius Ehret," *Proceedings of the Linnean Society*, 1894–95, 57–58.

22. See Natalie Rothstein, *Silk Designs of the Eighteenth Century in the Collection of the Victoria and Albert Museum, London* (London: Thames and Hudson, 1990), 33–48.

23. Letter from February 1741 to Mrs. Dewes, in *The Autobiography and Correspondence of Mary Glanville, Mrs. Delany*, ed. Lady Llanover, ser. 1 (London, 1861), 2:147–48.

24. Hannah More caricatured this fashion in the 1770s: "The other night we had

a great deal of company—eleven damsels to say nothing of men. I protest I hardly do them justice when I pronounce that they had, amongst them, on their heads, an acre and a half of shrubbery, besides slopes, grass plots, tulip beds, clumps of peonies, kitchen gardens, and greenhouses." Cited in Richard Corson, *Fashions in Hair: The First Five Thousand Years* (London: Owen, 1965), 348.

25. On Mary Delany see Ruth Hayden, *Mrs. Delany and Her Flower Collages* (London: British Museum Press, 1992), and Janice Farrar Thaddeus, "Mary Delany, Model to the Age," in *History, Gender, and Eighteenth-Century Literature*, ed. Beth Fowkes Tobin (Athens: U Georgia P, 1994).

26. Darwin, "Loves of Plants," 2:153–60.

27. *The Autobiography and Correspondence of . . . Mrs. Delany*, ed. Lady Llanover, ser. 2 (London, 1862), 3:95.

28. Cited in Hugh Curtis, *William Curtis, 1746–1799* (Winchester: Warren, 1941), 55.

29. Richard Mabey, *The Flowering of Kew: 350 Years of Flower Paintings from the Royal Botanic Gardens* (London: Century Hutchinson, 1988), 42. On Margaret Meen see also Henrey, *British Botanical and Horticultural Literature*, 2:248.

30. Henrey, *British Botanical and Horticultural Literature*, 2:580–81.

31. Mrs. Delany wrote in a letter dated August 16, 1772: "The Dss. of Portland has not this summer been edified by Mr. Lightfoot's lectures; he is sailing over lakes, traversing islands; clambering rocks, etc. etc., among the Western Isles of Scotland, in order to lay his prizes at her Grace's feet next Michaelmas" (*Autobiography and Correspondence*, ser. 2, 1:448). The duchess of Portland was also generous to Rousseau when he stayed at Wootten Hall for several months during 1766–67. Rousseau, having sought asylum in Switzerland after the publication and condemnation of *Emile* in 1762, then accepted an invitation to come to England. The duchess and Rousseau met through Mary Delany's brother, who lived near Rousseau's refuge at Wootten Hall in Derbyshire. She sent him plants and books during his stay, and he joined her on an expedition to the Peak District in search of wild plants in September 1766.

32. See Sylvia Harcstark Myers, *The Bluestocking Circle: Women, Friendship, and the Life of the Mind in Eighteenth-Century England* (Oxford: Clarendon, 1990), chap. 1 and passim.

33. Delany, *Autobiography and Correspondence*, ser. 2, 1:240.

34. Ibid., 238–39.

35. A. R. Horwood and C. W. F. Noel, *The Flora of Leicestershire and Rutland* (London: Oxford UP, 1933), cclxxii.

36. Letter to Sir Joseph Banks from J. Lloyd, dated November 15, 1779; in Dawson Turner Copies, Banks Correspondence, 1:273–74, Botany Library, Natural History Museum, London.

37. On Lady Anne Monson see James Britten, "Lady Anne Monson," *Journal of Botany* 56 (1918): 147–49. See also Ray Desmond, *The European Discovery of the Indian Flora* (Kew: Royal Botanic Garden; Oxford: Oxford UP, 1992), 149.

38. Cited in Blunt, *Compleat Naturalist*, 224.

39. A. A. Dallman and W. A. Lee, "An Old Cheshire Herbarium," *Lancashire and*

Cheshire Naturalist 10 (1917): 167. "The fact that this accomplished young lady incurred permanent invalidism and probably premature death, as a result of her activity as a botanist, gives a pathetic interest to the collection."

40. Within studies of art history, Ann Sutherland Harris and Linda Nochlin made this point in *Women Artists, 1550–1950* (New York: Knopf, 1977). On family patterns in music history, see Marcia J. Citron, "Women and the Lied, 1775–1850," in *Women Making Music: The Western Art Tradition*, ed. Jane Bowers and Judith Tick (Urbana: U of Illinois P, 1986).

41. B. D. Jackson, *Linnaeus* (London: Witherby, 1923), 318, 319. On Linnaeus and women, see Lisbet Koerner, "Women and Utility in Enlightenment Science," in *Configurations* 3, 2 (1995): 233–55.

42. Elisabeth Christina Linnea, "Om Indianska Krassens Blickande," *Svenska Kongliga Vetenskaps Academiens Handlingar* 23 (1762): 284–86.

43. Linnaeus appended to her account a description of three varieties of nasturtiums (of which one variety displays the particular phosphorescent effect), and a commentator from the Royal Society declares that this "remarkable observation . . . which has as a witness our greatest scientist cannot be doubted."

44. An early nineteenth-century researcher on women and the history of science and medicine claimed that she also wrote other botanical and phytological essays that were never published. Christian Friedrich Harless, *Die Verdienste der Frauen um Naturwissenschaft und Heilkunde* (Göttingen, 1830), 252–53.

45. Jackson, *Linnaeus*, 46.

46. Darwin's verse about the nasturtium (*Tropaeolum*) reads:

Ere the bright star, which leads the morning sky,
Hangs o'er the blushing east his diamond eye,
The chaste Tropaeo leaves her secret bed;
A faint-like glory trembles round her head;
Eight watchful swains along the lawns of night
With amorous steps pursue the virgin-light;
O'er her fair form the electric lustre plays,
And cold she moves amid the lambent blaze.

(4:43–50)

47. Letter dated October 1, 1755, cited in James Britten, "Jane Colden and the Flora of New York," *Journal of Botany* 33 (1895): 13.

48. *Jane Colden—Botanic Manuscript* (New York: Chanticleer, 1963). On Jane Colden see Marcia Myers Bonta, *Women in the Field: America's Pioneering Women Naturalists* (College Station: Texas A&M UP, 1991), 8.

49. On Ann Lee, see E. J. Willson, *James Lee and the Vineyard Nursery, Hammersmith* (London: Hammersmith Local History Group, 1961), 41–42. Also Patrick O'Brian, *Joseph Banks: A Life* (London: Collins Harvill, 1987), 151. On Ann Lee's drawings for Dr. Fothergill see *Journal of Botany* 52 (1914): 323. For examples of her work see Mabey, *Flowering of Kew*, passim.

50. James Lee sent Linnaeus a drawing by Ann Lee of the genus they thought an appropriate choice (Willson, *James Lee and the Vineyard Nursery*, 61).

51. On Anna Blackburne see V. P. Wystrach, "Anna Blackburne (1726–1793)—a Neglected Patron of Natural History," *Journal of the Society for the Bibliography of Natural History* 8 (1977): 148–68. On Forster and the Blackburnes see Michael E. Hoare, *The Tactless Philosopher: Johann Reinhold Forster, 1729–1798* (Melbourne: Hawthorn, 1976), 56–64.

52. *Bref och Skrifvelser af och till Carl von Linné* (Uppsala, 1916), 2, 1:285–86.

53. Wystrach, "Anna Blackburne," 153–54.

54. J. G. Dony, "Bedfordshire Naturalists: Charles Abbot (1761–1817)," *Bedfordshire Naturalist* 2 (1948): 39; also idem, *Flora of Bedfordshire* (Luton, 1953), 18–20. A writer remarked about Charles Abbot's *Flora Bedfordiensis* in the *Gentleman's Magazine* that "while [Abbot] wishes to promote the knowledge of botany among the ladies, he does not forget to introduce a most affectionate remembrance of his wife's talents in this line" (70 [1800]: 360).

55. Walter J. Ong, "Latin Language Study as a Renaissance Puberty Rite," *Studies in Philology* 56 (1959): 103–24.

56. *Gentleman's Magazine* 67, 1 (1797): 215.

57. Thomas Martyn, *Letters on the Elements of Botany*, 2d ed. (London, 1787), 12–13.

58. Ann Murry, *Sequel to Mentoria* (London, 1799), 250–51.

Chapter 3. Flora's Daughters as Writers

1. Janet Todd, ed., *A Dictionary of British and American Women Writers, 1660–1800* (Totowa, N.J.: Rowman and Allenheld, 1985), introduction. Judith Phillips Stanton, "Statistical Profile of Women Writing in English from 1660 to 1800," in *Eighteenth-Century Women and the Arts*, ed. Frederick M. Keener and Susan E. Lorsch (New York: Greenwood, 1988).

2. Janet Todd, *The Sign of Angellica: Women, Writing and Fiction, 1660–1800* (London: Virago, 1989), and Jane Spencer, *The Rise of the Woman Novelist: From Aphra Behn to Jane Austen* (Oxford: Blackwell, 1982).

3. Nancy Armstrong, *Desire and Domestic Fiction: A Political History of the Novel* (New York: Oxford UP, 1987); Spencer, *Rise of the Woman Novelist*; Mary Poovey, *The Proper Lady and the Woman Writer: Ideology as Style in the Works of Mary Wollstonecraft, Mary Shelley, and Jane Austen* (Chicago: U of Chicago P, 1984).

4. On the social construction of motherhood in eighteenth-century England, see Mitzi Myers, "Impeccable Governesses, Rational Dames, and Moral Mothers: Mary Wollstonecraft and the Female Tradition in Georgian Children's Books," *Children's Literature* 14 (1986): 31–59; Ruth Perry, "Colonizing the Breast: Sexuality and Maternity in Eighteenth-Century England," *Journal of the History of Sexuality* 2 (1991): 204–34; Elizabeth Kowaleski-Wallace, *Their Fathers' Daughters: Hannah More, Maria Edgeworth, and Patriarchal Complicity* (New York: Oxford UP, 1991); and Felicity A. Nussbaum, " 'Savage' Mothers: Narratives of Maternity in the Mid-Eighteenth Century," *Eighteenth-Century Life* 16 (February 1992): 163–84. On maternal ideology as context for poetry by male romantic writers, see Barbara Charlesworth Gelpi, *Shelley's Goddess: Maternity, Language, Subjectivity* (New York: Oxford UP, 1992).

5. "Over the fictional structures and topoi that make up Georgian juvenilia pre-

sides a maternal persona, controlling and dominating the tale in virtue's name" (Myers, "Impeccable Governesses," 34).

6. Frances Arabella Rowden, *A Poetical Introduction to the Study of Botany*, 3d ed. (London, 1818), vii–viii.

7. Ibid., ix–x.

8. Ibid., 130, 218–19, 214.

9. Darwin, "Loves of the Plants," 25–26. Darwin's personification of mimosa as female contrasts with the sexualized representation in James Perry's poem *Mimosa, or The Sensitive Plant* (London, 1779), where the plant is male and phallic. Perry's mimosa "enriched with sense and life; / Pleases the widow and the wife; / With its Articulations"; "We love to see it *rise* and *fall*, / And prove its innate *power*" (9, 5).

10. Rowden, *Poetical Introduction*, 248–51.

11. The coproprietor of the Hans Place School, Dominique de St. Quentin, a former ambassador from Louis XVI to the Court of St. James, wrote books based on the age-graded instructional methods of the French pedagogue Abbé Gaultier, among them *A Poetical Chronology of the Kings of England* (1792). (Gaultier developed a teaching method based on games, conversation, and synoptic tables.) St. Quentin may well have been the friend who suggested that Frances Rowden "compose a few elementary lessons on Botany, adapted to Abbé Gaultier's plan of instruction" (*Poetical Introduction* advertisement).

12. Mary Mitford, "On Revisiting the School Where I Was Educated," in *Poems* (London, 1810), and *Our Village*, in *The Works of Mary R. Mitford* (London, 1850), 111–16. The young Frances Kemble was a pupil at Rowden's school in Paris and wrote: "Miss Rowden's establishment in Hans Place had been famous for occasional dramatic representations by the pupils; and though she had become in her Paris days what in the religious jargon of that day was called serious, or even methodistical, she winked at, if she did not absolutely encourage, sundry attempts of a similar sort which her Paris pupils got up." Frances Kemble, *Record of a Girlhood* (London, 1878), 73, 109.

13. Rowden's other publications were *The Pleasure of Friendship* (1810; 3d ed., 1818); *A Christian Wreath for the Pagan Deities, or An Introduction to Greek and Roman Mythology* (1820); and *A Biographical Sketch of the Most Distinguished Writers of Ancient and Modern Times* (1821).

14. *Monthly Review* 70 (January 1813): 98–99; also *Critical Review*, ser. 4 (May 1812): 559. The *Anti-Jacobin Review* carried excerpts from her book and praised Rowden's design, particularly "her judgment in avoiding, as much as possible, the defects of her master [Erasmus Darwin]"; nevertheless, the reviewer wished "that our author had not taken Dr. Darwin for her model, who, amidst a croud [*sic*] of poetical beauties, exhibits so many meretricious embellishments, as tend to dazzle the eye and to bewilder the understanding" (10 [December 1801]: 356–67).

15. Murray, *British Garden*, 2d ed., preface. The author's name appeared on the title page of the second edition and remained there through later editions in 1808 and 1809. See Henrey, *British Botanical and Horticultural Literature*, 2:584, and G. E. Fussell, "The Rt. Hon. Lady Charlotte Murray," *Gardeners' Chronicle* 128 (1950): 238–39.

16. *Monthly Review* 31 (1800): 202. Likewise, the *British Critic* (15 [1800]: 86–87) referred to it as "a convenient enumeration" and noted its introduction "containing a familiar explanation of the Linnaean system."

17. Judith Pascoe, "Female Botanists and the Poetry of Charlotte Smith," in *Re-visioning Romanticism: British Women Writers, 1776–1837*, ed. Carol Shiner Wilson and Joel Haefner (Philadelphia: U of Pennsylvania P, 1994): "A new model of the Romantic poet is in order, one who, confronted with a field of daffodils, would count petals before launching into verse" (207). See also Stuart Curran, "Romantic Poetry: The I Altered," in *Romanticism and Feminism*, ed. Anne K. Mellor (Bloomington: Indiana UP, 1988).

18. Judith Phillips Stanton, "Charlotte Smith's 'Literary Business': Income, Patronage, and Indigence," in *The Age of Johnson: A Scholarly Annual*, ed. Paul Korshin (New York: AMS, 1987), 383–84.

19. Smith, *Rural Walks*, 2:125–28.

20. Smith, "The Goddess of Botany," lines 346–52, in *The Poems of Charlotte Smith*, ed. Stuart Curran (New York: Oxford UP, 1993), 231.

21. Letter dated March 15, 1798, in Smith, *Memoir and Correspondence*, 2:75.

22. Letter from Charlotte Smith to Thomas Caddell, Jr. and William Davies, written from Oxford and dated August 1, 1797 (Beinecke Library, Yale University). It will be published in Judith Phillips Stanton's forthcoming edition of the letters of Charlotte Smith.

23. Charlotte Smith, *Conversations Introducing Poetry: Chiefly on Subjects of Natural History* (London, 1804), 1:69, 71; 2:6–7.

24. "Flora" was later issued as part of Charlotte Smith's posthumous collection *Beachy Head, Fables, and Other Poems* (1807), "as many of her friends considered [the poems] as misplaced in [*Conversations for the Use of Children and Young Persons*], and not likely to fall under the general observation of those who were qualified to appreciate their superior elegance, and exquisite fancy." Advertisement, cited in *Poems of Charlotte Smith*, ed. Curran, 216.

25. Smith, *Conversations Introducing Poetry*, 2:174.

26. Florence M. A. Hilbish, "Charlotte Smith, Poet and Novelist (1749–1806)," (Ph.D. diss., University of Pennsylvania, 1941), 14–16.

27. Charlotte Smith, *The Young Philosopher* (London, 1798), 2:166.

28. On Henrietta Moriarty, see James Britten, "Mrs. Moriarty's 'Viridarium,'" *Journal of Botany* 55 (1917): 52–54.

29. Henrietta Moriarty, *Brighton in an Uproar* (London, 1811), 2:17–18, 82; 1:149.

30. Sarah Hoare, according to a Quaker memorial account, "loved the society of young people of all classes, to whom she was a very interesting companion, and she delighted to encourage the pursuit of knowledge which would improve the minds of her young associates." Her philanthropy was "proverbial." Active in issues of animal welfare, she wrote anguished personal memoranda about bull baiting and was instrumental in founding a "society for the promotion of humanity toward animals" in 1832. *Annual Monitor* (Society of Friends), 1856, 88–110.

31. Sarah Hoare, *Poems on Conchology and Botany* (London, 1831), 85.

32. Sarah Hoare, *A Poem on the Pleasures and Advantages of Botanical Pursuits* (London, 1826), viii–ix.

Chapter 4. Botanical Dialogues

1. See Gates, "Retelling the Story of Science"; Benjamin, "Elbow Room"; and Gates and Shteir, *Science in the Vernacular.*

2. Quantitative analysis of entries in R. B. Freeman's *British Natural History Books, 1495–1900: A Handlist* (Hamden, Conn.: Archon Books, 1980) shows that, among natural history books published during 1801–1900, 22 percent were on plants (cf. 14 percent on birds), and that women wrote a higher percentage of the botany books than of books about shells, insects, or birds (8 percent on botany, 2 percent on birds).

3. Books of this kind became "a steady part of women's publications after 1640." Patricia Crawford, "Women's Published Writings 1600–1700," in *Women in English Society, 1500–1800*, ed. Mary Prior (London: Methuen, 1985), 213. See also Hobby, *Virtue of Necessity*, chap. 7.

4. On the use of the dialogue form in newspaper advertisements for Packwood's razor strops during the 1790s, as part of the commercialization of shaving, see Neil McKendrick, John Brewer, and J. H. Plumb, *Birth of a Consumer Society: The Commercialization of Eighteenth-Century England* (London: Europa, 1982), 154–69.

5. Galileo, for example, wrote letters to convey his ideas to a more general readership; see, e.g., *Letters on Sunspots* and his 1615 letter to the Grand Duchess Christina. For five decades beginning in the late seventeenth century, Antonin van Leeuwenhoek used letters to communicate his research findings about microorganisms to private scientists and also to the Royal Society of London. Robert Boyle chose to publish accounts of several of his experiments in letter form specifically in order to proselytize for his approach by encouraging correspondents to try to replicate his findings; Steven Shapin and Simon Schaffer have argued that Boyle, by sending epistolary protocols to nephews and friends, sought to multiply witnesses for his experiments and for his experimental programme (*Leviathan and the Air-Pump*, 59).

6. Letter to William Kirby, January 25, 1806, in John Freeman, *Life of the Rev. William Kirby, M.A., Rector of Barham* (London: Longman, 1852), 286–90.

7. Aphra Behn translated Fontenelle's dialogue as *The Theory or System of Several New Inhabited Worlds* (1688), and Elizabeth Carter translated Algarotti's dialogue as *Sir Isaac Newton's Philosophy Explained for the Use of the Ladies* (1739). See Fontenelle's *Conversations on the Plurality of Worlds*, trans. H. A. Hargreaves, with introduction by Nina R. Gelbart (Berkeley and Los Angeles: U of California P, 1990). On traditional uses of dialogues for literature, philosophy, science, and pedagogy in the Renaissance, see Virginia Cox, *The Renaissance Dialogue: Literary Dialogue in Its Social and Political Contexts, Castiglione to Galileo* (Cambridge: Cambridge UP, 1992).

8. Hester Chapone, "On Conversation," in *Miscellanies in Prose and Verse* (London, 1775), 40.

9. See Peter Burke, *The Art of Conversation* (Ithaca: Cornell UP, 1993), esp. 108–17.

10. Maria Edgeworth, *Practical Education* (London, 1798), 1:vi. She also wrote: "From conversation, if properly managed, children may learn with ease, expedition,

and delight . . . ; and a skilful preceptor can apply in conversation all the principles that we have laboriously endeavoured to make intelligible in a quarto volume" (appendix).

11. Priscilla Wakefield, *An Introduction to Botany in a Series of Familiar Letters* (London: Newbery, 1796), 27, 69.

12. Marina Benjamin suggests that Priscilla Wakefield may have used natural theology as "a ploy to circumvent censorship": "Advocating a conservative natural theology was expedient for women of a liberal persuasion, who, responding to public pressure, felt a need to distinguish their own liberality from that of the republicans" ("Elbow Room," 39).

13. During twenty-two years as a publisher, Elizabeth Newbery issued over three hundred books for "children and young people." See S. Roscoe, *John Newbery and His Successors, 1740–1814: A Bibliography* (Wormley: Five Owls, 1973). On the publishing firm of Darton and Harvey, see Linda David, *Children's Books Published by William Darton and His Sons* (Bloomington: Lilly Library, Indiana U, 1992).

14. Books by Priscilla Wakefield include *Domestic Recreations, or Dialogues Illustrative of Natural and Scientific Subjects* (1805), and *The Juvenile Travellers* (1801).

15. *Gentleman's Magazine* 87, 2 (1817): 54.

16. Sarah Atkins Wilson, *Botanical Rambles* (London, 1822), 32.

17. For discussion of representations of women's fictional letters that move beyond canonical epistolary novels, see Mary A. Favret, *Romantic Correspondence: Women, Politics and the Fiction of Letters* (Cambridge: Cambridge UP, 1993).

18. *New British Lady's Magazine*, August 1817, 127.

19. Myers, "Fictions for Facts," 233. In its written form the dialogue is a fictional conversation that can serve as a debate, a conference, a Socratic dialogue, a drama. It can be an elaborated monologue masking as a familiar conversation; in Sarah Trimmer's *Easy Introduction to the Study of Nature* (1780), for example, the narrator talks to children as they walk together, but the children have no narrative opportunity to ask questions. See Myers, "Science for Women and Children." Also Marion Amies, "Amusing and Instructive Conversations: The Literary Genre and Its Relevance to Home Education," *History of Education* 14, 2 (1985): 87–99.

20. Sarah Fitton, *Conversations on Botany*, 2d ed. (London: Longman, 1818), 121.

21. Ibid., 44.

22. William Fitton (1780–1861) was a fellow of the Royal Society, the Linnean Society, the Astronomical Society, and the Royal Geographical Society. During the years 1827–46 Fitton held office as president and vice president of the Geological Society. "Obituary Notice of Dr. Fitton," *Quarterly Journal of the Geological Society of London* 18 (1862): xxx–xxxiv.

23. D. J. Mabberly, *Jupiter Botanicus: Robert Brown of the British Museum* (London: British Museum [Natural History], 1985), 243, 343.

24. One story, *How I Became a Governess* (1861), tells of a girl from a good family background, sent away to school at age seventeen, who, on her own in the world, sought a job as a governess. Needing French for the job, she went to France to work as an English teacher while learning French. Descriptions in the story of the teacher's life

in a school with penurious accommodations have a specificity that suggests firsthand experience.

25. Harriet Beaufort, *Dialogues on Botany* (London: Hunter, 1819), v.

26. The correspondence of Maria Edgeworth offers occasional glimpses of Harriet Beaufort. The charm and learning of the mathematician Mary Somerville, she wrote, "puts me in mind of Harriet Beaufort more than any one I ever saw." Letter to Mrs. Edgeworth, January 16, 1822, in *Maria Edgeworth: Letters from England, 1813–1844*, ed. Christina Colvin (Oxford: Clarendon, 1971).

27. C. C. Ellison, *The Hopeful Traveller: The Life and Times of Daniel Augustus Beaufort LL.D., 1739–1821* (Kilkenny: Boethius, 1987), 78. See also Alfred Friendly, *Beaufort of the Admiralty: The Life of Francis Beaufort, 1774–1857* (London: Hutchinson, 1977).

28. See M. E. Mitchell, "The Authorship of *Dialogues on Botany,*" *Irish Naturalists' Journal* 19, 11 (1979): 407.

29. Beaufort, *Dialogues on Botany*, 125, 158.

30. *New Monthly Magazine* 69 (1819): 330; *Monthly Review* 92 (1820): 439–42.

31. Mary Roberts, *Flowers of the Matin and Even Song, or Thoughts for Those Who Rise Early* (London: Grant and Griffiths, 1845), 99–100.

32. For a recuperative view of Mary Roberts, with special focus on her *Conchologist's Companion*, see Stephen Jay Gould, "The Invisible Woman," *Natural History* 102 (June 1993): 14–23.

33. Mary Roberts, *Wonders of the Vegetable Kingdom Displayed* (London: Whittaker, 1822), 72.

34. Nicola J. Watson, *Revolution and the Form of the British Novel, 1790–1825: Intercepted Letters, Interrupted Seductions* (Oxford: Clarendon, 1994).

35. Roberts, *Wonders of the Vegetable Kingdom*, 49–50.

36. *Eclectic Review* (1823), n.s., 18 (1823): 561.

37. Roberts, *Wonders of the Vegetable Kingdom*, 38–39.

38. Mary Roberts, *Annals of My Village* (London, 1831), iv.

39. *Conversations on Chemistry* moved through sixteen editions over the next five decades in England. It was also much reprinted in America; see M. Susan Lindee, "The American Career of Jane Marcet's *Conversations on Chemistry,* 1806–1853," *Isis* 82 (1991): 8–23. Other well-known titles by Jane Marcet were *Conversations on Political Economy* (1817) and *Conversations on Natural Philosophy* (1820).

40. Jane Marcet, *Conversations on Vegetable Physiology* (London: Longman, 1819), 1:27.

41. Elizabeth Chambers Patterson, *Mary Somerville and the Cultivation of Science, 1815–40* (Boston: Nijhoff, 1983), 10–11; and Edgeworth, *Letters from England, 1813–1844*, letter dated January 16, 1822.

42. Cited in *Maria Edgeworth in France and Switzerland: Selections from the Edgeworth Family Letters*, ed. Christina Colvin (New York: Oxford UP, 1979), xxiii. On Restoration Geneva and opportunities for women and science, see Clarissa Campbell Orr, "Albertine Necker de Saussure, the Mature Woman Author, and the Scientific Education of Women," *Women's Writing: The Elizabethan to Victorian Period* 2 (1995).

43. Genevans paid to attend Candolle's public botanical lectures during 1816–28, with funds going partially to developing the Museum of Natural History. His lectures, he wrote, were "fréquenté par toute la première société de la ville . . . et mit la botanique un peu à la mode." Alphonse de Candolle, ed., *Mémoires et souvenirs de Augustin-Pyramus de Candolle* (Geneva, 1862), 287, 330.

44. Elizabeth Chambers Patterson, *Mary Somerville and the Cultivation of Science, 1815–1840* (Boston: Nijhoff, 1983), 146; Harriet Martineau, *Biographical Sketches* (New York: Hurst, 1868), 72.

45. Edgeworth, *Letters from England, 1813–1844*, letter dated January 5, 1822.

46. *Magazine of Natural History* 50 (1830): 147; 2, 11 (1829): 454.

47. The principal proponent of this view is Kowaleski-Wallace in *Their Fathers' Daughters*.

48. Armstrong, *Desire and Domestic Fiction*.

Chapter 5. Three "Careers" in Botanical Writing

1. Maria Jacson, *Botanical Dialogues, between Hortensia and Her Four Children* (London: J. Johnson, 1797), 3–4.

2. Ibid., 72, 229–30.

3. Ibid., 153–54.

4. Ibid., 53–55.

5. A commendatory letter written by Erasmus Darwin (dated August 24, 1795) and printed with the book read:

> According to your desire, Sir Brooke Boothby and myself have been agreeably busied for many days in reading and considering your Botanical Dialogues for Children; and much admire your address in so accurately explaining a difficult science in an easy and familiar manner, adapted to the capacities of those, for whom you professedly write; and at the same time making it a compleat elementary system for the instruction of those of more advanced life, who wish to enter upon this entertaining, though intricate study. We think therefore, that not only the youth of both sexes, but the adults also, will be much indebted to your ingenious labours, which we hope you will soon give to the public. (King-Hele, *Letters of Erasmus Darwin*, 287–88).

6. Maria Jacson, *Botanical Lectures* (London: J. Johnson, 1804), iii.

7. *Analytical Review* 28 (1798): 398.

8. Maria Jacson, *Sketches of the Physiology of Vegetable Life* (London: Colburn, 1811), 27.

9. Ibid., 126.

10. An early general essay, "Mrs. Maria Elizabeth Jacson," by G. E. Fussell appeared as part of a series "Some Lady Botanists of the Nineteenth Century," in the *Gardeners' Chronicle* 130 (1951): 63–64. See also Henrey, *British Botanical and Horticultural Literature*, 2: 581–84. (Henrey misidentifies Maria Jacson as the wife of Roger Jacson; Roger Jacson was Maria Jacson's older brother.) Desmond King-Hele supplied further key details in *Letters of Erasmus Darwin*, 287–88, nn. 1–3. The National Union Catalogue incorrectly lists Maria Elizabeth Jacson as author of *The Pictorial*

Flora (1840); this was written by Mary Anne Jackson (fl. 1830s–1840s). Diaries in the Lancashire Record Office, dated 1829–37, written by her older sister, Frances Jacson, contain details about family and friends (DX 267–78). See also Joan Percy, "Maria Elizabetha Jacson and her *Florist's Manual,*" *Garden History* 20 (1992): 45–56.

11. Letter dated November 24, 1818, from Byrkley Lodge, near Lichfield. *Maria Edgeworth: Letters from England*, 141. Frances Jacson's two published novels were *Rhoda* (ca. 1813) and *Things by Their Right Names* (1812).

12. Maria Jacson, *A Florist's Manual: Hints for the Construction of a Gay Flower-Garden* (London: Colburn, 1816), 3.

13. Darwin, *Botanic Garden*, part 2, "Supplement," to "The Loves of the Plants," "Addition to the Note on Silene," 181. In *Zoonomia; or the Laws of Organic Life* he wrote: "Miss M. E. Jacson acquainted me, that she witnessed this autumn an agreeable instance of sagacity in a little bird, which seemed to use the means to obtain an end, the bird repeatedly hopped upon a poppy-stem and shook the head with its bill, till many seeds were scattered, then it settled on the ground and eat the seeds, and again repeated the same management. Sept. 1, 1794" (2d ed., 1796, 1:161).

14. Erasmus Darwin, *A Plan for the Conduct of Female Education in Boarding Schools* (1797; rpt. New York: Johnson, 1968), 41.

15. Reverend Simon Jacson made his will in February 1796 and again in April 1799. A codicil, proved in 1815, added a small annual sum to be paid jointly to Maria and her sister (Cheshire Record Office, Chester, DDX 9/19).

16. *Monthly Review*, enl. ser., 24 (1797): 449–50. The *Critical Review* commended the author for her industry on behalf of the science of botany, especially for "the familiarity of [her] manner, and the clearness of her descriptions" in explaining the Linnaean language of botany, but felt called on to position the praise by declaring that *Botanical Dialogues* is a book for young students, and hence "those who have made some proficiency in the study, will not be equally pleased with the work, as it is not (and indeed was not intended to be) profoundly scientific" (2d ser., 22 [1798]: 446–49).

17. Edgeworth, *Practical Education*, 1:366.

18. Jacson, *Botanical Dialogues*, 238–39.

19. Darwin, *Plan for the Conduct of Female Education*, 10.

20. *Monthly Review*, enl. ser., 65 (May–August 1811): 108.

21. See Poovey, *Proper Lady and the Woman Writer*.

22. Jacson, *Florist's Manual*, 12. See Percy, "Maria Elizabetha Jacson."

23. John Sutherland, "Henry Colburn, Publisher," *Publishing History* 19 (1986): 59–84.

24. Jane Loudon's remark is cited in Bea Howe, *Lady with Green Fingers: The Life of Jane Loudon* (London: Country Life, 1961), 47–48.

25. Articles by Agnes Ibbetson appeared in William Nicholson's *Journal of Natural Philosophy, Chemistry, and the Arts* vols. 23–36 (1809–13), passim; the *Philosophical Magazine* vols. 43–60 (1814–22), passim; and *Annals of Philosophy* vols. 11–14 (1818–19), passim. Translated extracts from her essays appeared in *Journal de Botanique Appliquée* (Paris), *Annali di Scienza e Lettere* (Milan), and *Bibliothèque Britannique*. The last journal, published in Geneva, carried monthly reports about developments

in British science during the years of the Napoleonic Wars, when contact was curtailed between England and continental Europe.

26. *Curtis's Botanical Magazine* 31 (1810), pl. 1259. On Agnes Ibbetson see also *Gentleman's Magazine* 93, 1 (1823): 474, and *Dictionary of National Biography*, 1921.

27. There are five manuscript volumes of writing and drawing by Agnes Ibbetson in the Botany Library of the Natural History Museum, London: three folio volumes of watercolor drawings of grasses, with notes; a notebook on botanical and other subjects, illustrated with pencil drawings of plants and working drawings of plant structure, and some pages of personal journal entries; and a "Botanical Treatise" in the form of letters (MSS IBB). The Linnean Society, London, holds a manuscript treatise, "Phytology," and letters to Sir James E. Smith (MSS 120a, 120b, 490). It has not proved possible to trace a collection of wood samples listed in the catalog of the Natural History Museum, London.

28. François Delaporte rehearses debates about "animality" and "vegetality" in *Nature's Second Kingdom*, trans. Arthur Goldhammer (Cambridge: MIT P, 1982).

29. Nicholson's *Journal* 24 (1809): 114.

30. *Philosophical Magazine* 45 (1815): 184.

31. Cited in Rev. William Webb, *Memorials of Exmouth* (1872), 53–54. Several members of the Scottish Thomson family, of whom Thomas Thomson was a part, were active in botanical work. Agnes Ibbetson's husband, James Ibbetson, eldest son of the archdeacon of St. Albans, wrote essays in legal history. I acknowledge the assistance of Guy Holborn, librarian of Lincoln's Inn, in tracing information about James Ibbetson.

32. Botany Library, Natural History Museum, London, MSS IBB. The ten pages of journal entries cover a two-week period in December of an unspecified year, but internal references place it in 1803–8.

33. Barbara Stafford, *Body Criticism: Imaging the Unseen in Enlightenment Art and Medicine* (Cambridge: MIT P, 1991); Meyer, *Scientific Lady in England*; and Phillips, *Scientific Lady*, 148–50.

34. On eighteenth-century microscopy, see Brian J. Ford, *Single Lens: The Story of the Simple Microscope* (New York: Harper and Row, 1985), chap. 7; Brian Bracegirdle, *A History of Microtechnique* (Ithaca: Cornell UP, 1978), chap. 2; and S. Bradbury, "The Quality of the Image Produced by the Compound Microscope: 1700–1840," in *Historical Aspects of Microscopy*, ed. S. Bradbury and G. L'E. Turner (Cambridge: Heffner, 1967).

35. Agnes Ibbetson, "On the Interior Buds of All Plants," Nicholson's *Journal* 33 (1812): 10.

36. *Philosophical Magazine* 48 (1816): 97.

37. Nicholson's *Journal* 35 (1813): 88; 33 (1812): 243; 33 (1812): 10, 177.

38. See Evelyn Fox Keller, *A Feeling for the Organism: The Life and Work of Barbara McClintock* (New York: W. H. Freeman, 1983). Agnes Ibbetson's approach to botanical research also anticipates the work in our day of the developmental biologist Dr. Victoria Elizabeth Foe, who combines observation over long periods with microscopic study of embryonic growth, illustrating her findings with detailed drawings. *New York Times*, August 10, 1993, B5.

39. Nicholson's *Journal* 28 (1811): 98–103.

40. She wrote: "If this matter was to be decided by argument or general knowledge, I should never attempt to contradict a gentleman of Mr. Knight's understanding and abilities; but, after the quantities of dissections I have made (and I have many hundred barks drawn by me, besides the twenty I have taken to pieces) I must be well acquainted with that part, and there certainly is not a single sap-vessel to be found in it, except the nourishing vessels passing in their separate cylinders to the leaves." Nicholson's *Journal* 35 (1813): 94.

41. *Philosophical Magazine* 48 (1816): 283.

42. Nicholson's *Journal* 23 (1809): 300.

43. *Monthly Magazine* 31 (1811): 601.

44. Brian Ford, author of *Images of Science: A History of Scientific Illustration* (London: British Library, 1992), personal communication.

45. "Botanical Treatise," Botany Library, Natural History Museum, London (MSS IBB).

46. *Philosophical Magazine* 48 (1816): 97.

47. Letter, May 5, 1814 (Linnean Society, MS 120a; see also MSS 120b, 490).

48. "Phytology," 45–46 (Linnean Society MS 120a).

49. *Philosophical Magazine* 56 (1820): 4.

50. *Philosophical Magazine* 64 (1824): 81.

51. She wrote: "He is also of my opinion respecting the supposed circulation of sap, which he appears persuaded does not exist." Nicholson's *Journal* 31 (1812): 162.

52. *Philosophical Magazine* 59 (1822): 244.

53. Agnes Ibbetson, "On the Adapting of Plants to the Soil, and Not the Soil to the Plants," *Letters and Papers on Agriculture, Planting, etc. Selected from the Correspondence of the Bath and West of England Society* 14 (1816): 136–59. Ibbetson appears as an honorary and corresponding member in the *Rules, Orders, and Premiums of the Bath and West of England Society for the Year 1814* (Bath, 1814), the only woman on that list.

54. Her articles "On the Death of Plants," "On the Injurious Effects of Burying Weeds," and "On the Action of Lime on Animal and Vegetable Substances" appeared in *Annals of Philosophy* 11 (1818): 252–62; 12 (1818): 87–91; and 14 (1819): 125–29.

55. *Annals of Philosophy* 14 (1819): 125.

56. Botany Library, Natural History Museum, London. MSS IBB, letter 7.

57. When the botanist John Bostock met her in 1814 and wrote to Smith on her behalf, he himself had not yet been elected to fellowship in the Linnean Society. The only other scientist of standing to whom Ibbetson refers, Sir William Herschel, visited her about 1816 and witnessed some of her findings, but no details of that relationship are known.

58. See Anne K. Mellor, *Romanticism and Gender* (New York: Routledge, 1993), passim. See also Stuart Curran, "Romantic Poetry," in Mellor, *Romanticism and Feminism*, 189–90.

59. See *The Letters of Mary Wollstonecraft Shelley*, ed. Betty T. Bennett (Baltimore: Johns Hopkins UP, 1980), passim, and *The Letters of Percy Bysshe Shelley*, ed. Frederick L. Jones (Oxford: Clarendon, 1964), esp. vol. 2 passim.

60. On Elizabeth Kent see Molly Tatchell, *Leigh Hunt and His Family in Hammer-*

smith (London: Hammersmith Local History Group, 1969). Leigh Hunt wrote this sonnet "To Miss K.":

There, Bess, your namesake held not sceptred hand
Under a canopy, so full and bright,
Not even that which Spenser hung with light,
And little shouldering angels made expand,
When she sat arbitress of fairy-land.
Fancy a sun o'er head, to make the sight
Warm outwards, and a bank with daisies white,
And you're a rural queen, finished and fanned.
And now what sylvan homage would it please
Your Leafyship to have? bracelets of berries,
Feathers of jays, or tassels made of cherries,
Strawberries and milk, or pippins crisp to squeeze?
No, says your smile, — but two things richer far
A verse, and a staunch friend; — and here they are.

Leigh Hunt, *Foliage, or Poems Original and Translated* (London, 1818)

Leigh Hunt's many letters to "Bessy" Kent from Italy in the early and mid-1820s chronicle a complex attachment that led at one point to strained relations between the sisters and to concerns about "calumny" and reputation. *The Correspondence of Leigh Hunt*, 2 vols. (London: Smith and Elder, 1862), passim.

61. *Archives of the Royal Literary Fund*, file 1383 (applications dating May 25, 1855, May 2, 1858, and December 1, 1860, microfilm). On the Royal Literary Fund, founded in 1788, whose beneficiaries had included Charlotte Lennox, Samuel Coleridge, and Leigh Hunt, see Nigel Cross, *The Royal Literary Fund, 1790–1918: An Introduction to the Fund's History and Archives, with an Index of Applicants* (London: World Microfilm, 1984).

62. Dan Cruickshank and Neil Burton, *Life in the Georgian City* (London: Viking, 1990), 194–95; and Bewell, "Keats's 'Realm of Flora,'" 76–77.

63. Elizabeth Kent, *Flora Domestica, or The Portable Flower-Garden* (London, 1823), xiii.

64. Royal Literary Fund, file 1383, doc. 20.

65. Elizabeth Kent, *Sylvan Sketches, or A Companion to the Park and Shrubbery* (London: Taylor and Hessey, 1825), preface, 60.

66. Letter dated August 16, 1823, in *The Collected Letters of Samuel Taylor Coleridge*, ed. E. L. Griggs (Oxford: Clarendon, 1971), 5:1344–45.

67. Letters dated July 31, 1823, and August 1823, in *The Letters of John Clare*, ed. Mark Storey (Oxford: Clarendon, 1985). Clare also sent an eight-page letter to his publisher about *Flora Domestica* with "notes and remarks" about individual plants and flowers in Kent's book; "I never," he wrote, "met with a book about flowers that pleasd me so much in my life." See *The Natural History Prose Writings of John Clare*, ed. Margaret Grainger (Oxford: Clarendon, 1983), 10–25.

68. Letter dated April 8, 1825, in *Letters of Mary Wollstonecraft Shelley*, 1:477.

69. *Monthly Review* 104 (1824): 155–56.

70. *Magazine of Natural History* 3 (1830): 57.

71. Ibid., 1 (1828): 229.

72. Ibid., 1:129.

73. Ibid., 128.

74. Ibid., 126.

75. Letter to John Hamilton Reynolds, dated September 21, 1817, in *The Letters of John Keats*, ed. Maurice Buxton Forman (London: Oxford UP, 1952), 44–45. Thomas Moore satirized a woman with scientific interests in the comic opera *M. P., or The Bluestocking* (1811). See Myers, *Bluestocking Circle*, epilogue, "The Term 'Bluestocking' and the Bluestocking Legacy."

76. *Examiner*, no. 881, December 19, 1824, 303–4.

77. See Alan Richardson, "Romanticism and the Colonization of the Feminine," in Mellor, *Romanticism and Feminism*.

78. On Hunt's contributions to her writing, see Molly Tatchell, "Elizabeth Kent and *Flora Domestica*," *Keats-Shelley Memorial Bulletin* 27 (1976): 15–18. See also Hunt, *Correspondence*, 1:204–6, 216–18.

79. Letter dated September 7, 1824, Hunt, *Correspondence*, 1:234.

80. Letters dated May 5, 1825, and April 11, 1826, in *Letters of John Clare*.

81. "I had long been collecting materials for a third volume upon *wild flowers*, when I had the gratification of reading a letter addressed to my publisher by Coleridge the poet, approving my first book [*Flora Domestica*], and recommending them to *engage* me to write a similar one upon the wild-flowers; which they were not aware that I had commenced" (RLF, 1383, doc. 20).

82. *Gardener's Magazine* 3 (1828): 104.

83. *Archives of the Royal Literary Fund*, 1383. Letter dated May 24, 1855.

84. See S. D. Mumm, "Writing for Their Lives: Women Applicants to the Royal Literary Fund, 1840–1880," *Publishing History* 27 (1990): 27–47.

85. *Magazine of Natural History* 1 (1828): 133–34.

Chapter 6. Defeminizing the Science of Botany

1. Sarah Waring, *A Sketch of the Life of Linnaeus* (London: Harvey and Darton, 1827), 6–10.

2. Geoffrey Cantor, *Michael Faraday: Sandemanian and Scientist* (London: Macmillan, 1991), 126.

3. Iwan Morus, Simon Schaffer, and Jim Secord, "Scientific London," in *London, World City, 1800–1840*, ed. Celina Fox (New Haven: Yale UP, 1992), 129–42.

4. *Literary Gazette*, no. 1911 (September 3, 1853).

5. Steven Shapin, "'A Scholar and a Gentleman': The Problematic Identity of the Scientific Practitioner in Early Modern England," *History of Science* 29 (1991): 313.

6. Boyd Hilton, *The Age of Atonement: The Influence of Evangelicalism on Social and Economic Thought, 1795–1865* (Oxford: Clarendon, 1988), chap. 1.

7. Peter Bailey, *Leisure and Class in Victorian England* (London: Methuen, 1987), 77.

8. Desmond, *Celebration of Flowers*, 121.

9. See Anne Secord, "Science in the Pub: Artisan Botanists in Early Nineteenth-

Century Lancashire," *History of Science* 32 (1994): 269–315. See also "Botanists of Manchester," *Chambers's Journal* 30 (1858): 255–56.

10. On the field club tradition, see Allen, *Naturalist in Britain*, chap. 8.

11. David Elliston Allen, *The Victorian Fern Craze: A History of Pteridomania* (London: Hutchinson, 1969).

12. Allen, *Naturalist in Britain*, 137.

13. *Gardeners' Chronicle* 158 (1965): 434. On John Lindley, see Frederick Keeble, "John Lindley, 1799–1865," in Oliver, *Makers of British Botany; Journal of Botany* 3 (1865): 384–88; William Gardener, "John Lindley," *Gardeners' Chronicle* 158 (1965): 386 passim; and *Dictionary of Scientific Biography*, 1973, 8:371–73. A full biography of John Lindley is much needed.

14. On technical matters in early nineteenth-century botany, see Morton, *History of Botanical Science*, chap. 9, and Peter F. Stevens, *The Development of Biological Systematics: Antoine-Laurent de Jussieu, Nature, and the Natural System* (New York: Columbia UP, 1994). On Robert Brown, called "botanicorum facile princeps," see Mabberly, *Jupiter Botanicus*.

15. Lindley, *Introductory Lecture Delivered in the University of London*, 9–10.

16. Adrian Desmond, *The Politics of Evolution: Morphology, Medicine, and Reform in Radical London* (Chicago: U of Chicago P, 1989), 26–27.

17. Frederick Keeble, "John Lindley, 1799–1865," in Oliver, *Makers of British Botany*, 167.

18. Lindley, *Introductory Lecture*, 14.

19. Ibid., 17.

20. In her study of working-class botany, Anne Secord makes the point that not only women but also workingmen were being excluded; see "Science in the Pub."

21. See, for example, Philippa Levine, *The Amateur and the Professional: Antiquarians, Historians and Archaeologists in Victorian England, 1838–1886* (Cambridge: Cambridge UP, 1986). On disciplinary knowledge production, including construction of fields and socializing practices, see Ellen Messer-Davidow, David R. Shumway, and David J. Sylvan, eds., *Knowledges: Historical and Critical Studies in Disciplinarity* (Charlottesville: UP of Virginia, 1993).

22. See Robert M. Young, "The Historiographical and Ideological Contexts of the Nineteenth-Century Debate on Man's Place in Nature," in *Darwin's Metaphor: Nature's Place in Victorian Culture* (Cambridge: Cambridge UP, 1985); and Frank M. Turner, *Contesting Cultural Authority: Essays in Victorian Intellectual Life* (Cambridge: Cambridge UP, 1993), chap. 7.

23. See Anita Levy, *Other Women: The Writing of Class, Race, and Gender, 1832–1898* (Princeton: Princeton UP, 1991), esp. chap. 1.

24. Charlotte Elizabeth Tonna, *Chapters on Flowers*, 3d ed. (London, 1839), 2, 97.

25. On the language of flowers tradition in England, see Beverly Seaton, "Considering the Lilies: Ruskin's 'Proserpina' and Other Victorian Flower Books," *Victorian Studies* 28, 2 (1985): 256; and Michael Waters, *The Garden in Victorian Literature* (Aldershot: Scolar, 1988), chap. 6. See also Goody, *Culture of Flowers*.

26. See Ray Desmond, "Victorian Gardening Magazines," *Garden History* 5, 3 (1977): 47–66.

27. *Botanist* 1 (1836), preface.

28. *Phytologist* 1 (1841), preface.

29. G. Francis, *The Grammar of Botany* (London, 1840), preface.

30. *Letters of Euler on Different Subjects in Natural Philosophy. Addressed to a German Princess*, ed. David Brewster (New York, 1833), 1:11–12.

31. J. S. Forsyth, *The First Lines of Botany, or Primer to the Linnaean System* (London: James Bulcock, 1827), 17.

32. Forsyth's titles include *The Mother's Medical Pocket Book* (1824), *The New London Medical and Surgical Dictionary* (1826), and *Demonologia* (1827). *Demonologia*, an "Exposé of Ancient and Modern Superstitions," contrasts "the stench and vapour of the train oil of ignorance and superstition, . . . with the brilliant gas of reason and comparative understanding" (x). John Symons, chief cataloger, Wellcome Institute for the History of Medicine, generously provided information and materials. Applications by John S. Forsyth to the Royal Literary Fund (file 900) during 1837–40 contain documentation about his career as author of more than thirty books issued by at least twenty-five different booksellers.

33. Cited in Gardener, "John Lindley," 409.

34. John Lindley, *Introduction to Botany* (London, 1832), vii; idem, *School Botany*, 12th ed. (London, 1862), iv.

35. Charles Knight, *Passages of a Working Life during Half a Century* (London, 1864), 2:236.

36. John Lindley, *Ladies' Botany, or A Familiar Introduction to the Study of the Natural System of Botany* (London: Ridgway, 1834–37), 2:2; 1:iv.

37. Richard Yeo, "Science and the Organization of Knowledge in British Dictionaries of Arts and Sciences, 1730–1850," *Isis* 82, 311 (1991): 43–48.

38. See Gaye Tuchman, with Nina E. Fortin, *Edging Women Out: Victorian Novelists, Publishers, and Social Change* (New Haven: Yale UP, 1989).

39. Myers, "Science for Women and Children," 181.

40. Richmal Mangnall, *Historical and Miscellaneous Questions for the Use of Young People* (London, 1800), preface.

41. Because the catechism was identified with traditional patterns of authority and religious obedience and thus with narratively embodied conservative social and political beliefs, progressive educators in the late eighteenth century opposed it, explicitly on pedagogical grounds but implicitly for political reasons. Rousseau criticized it in his discussions of religious education in *Emile* for boys and also for girls. In *Practical Education* Maria Edgeworth and Richard Lovell Edgeworth objected to a history book written in catechistic form that starts with abstract concepts rather than the world of experience. They acknowledge that children can learn to answer "a metaphysical catechism," but they ask, "When children are perfect in their responses to all those questions, how much are they advanced in real knowledge?" There was considerable discussion around the turn into the nineteenth century between those who did and did not like science books that taught "matters of fact." Whereas one reviewer commended the teaching style of Samuel Parkes's *Chemical Catechism* (London, 1808) as "the best mode for conveying instruction," another criticized this format as "likely to encumber the young mind with very imperfect outlines of various

abstruse, and probably to him utterly unintelligible subjects." These differences in opinion were partly about pedagogy and partly about politics.

42. Daniel Oliver, *Lessons in Elementary Botany: The Part of Systematic Botany Based upon Material Left in Manuscript by the Late Professor Henslow* (London: Macmillan, 1864), vi.

43. See David Layton, *Science for the People: The Origins of the School Science Curriculum in England* (New York: Science History Publications, 1973), chap. 3; and Jean P. Bremner, "Some Aspects of Botany Teaching in English Schools in the Second Half of the Nineteenth Century," *School Science Review* 38 (1956–57): 376–83.

44. Allen, *Naturalist in Britain*, 184.

Chapter 7. Women and Botany in the Victorian Breakfast Room

1. Leonore Davidoff and Catherine Hall, *Family Fortunes: Men and Women of the English Middle Class, 1780–1850* (Chicago: U of Chicago P, 1987), chaps. 1–3.

2. *The Young Lady's Book of Botany* (London, 1838), preface.

3. See F. K. Prochaska, *Women and Philanthropy in Nineteenth-Century England* (Oxford: Clarendon, 1980); Moira Ferguson, *Subject to Others: British Women Writers and Colonial Slavery, 1670–1834* (New York: Routledge, 1992), 258–64.

4. See Secord, "Science in the Pub." J. S. Henslow included the daughters of agricultural laborers in the botany classes he gave at Hitchin during the 1850s.

5. In the opening years of the Royal Institution, founded in 1799 "to direct the public attention to the arts, by an establishment for diffusing the knowledge and facilitating the general introduction of useful mechanical inventions and improvements," women not only attended general lectures, but also were allowed to join as subscribers in their own right. Gwendy Caroe, *The Royal Institution: An Informal History* (London: John Murray, 1985), 16. See also George A. Foote, "Sir Humphry Davy and His Audience at the Royal Institution," *Isis* 43 (1952): 6–12.

6. Jack Morrell and Arnold Thackray, *Gentlemen of Science: Early Years of the British Association for the Advancement of Science* (Oxford: Clarendon, 1981), 148–57.

7. David E. Allen, "The Women Members of the Botanical Society of London, 1836–56," *British Journal for the History of Science* 13, 45 (1980): 240–54; see also idem, *The Botanists: A History of the Botanical Society of the British Isles through 150 Years* (Winchester: St. Paul's Bibliographies, 1986).

8. *Ladies' Companion* 1 (1849): 8. On elaborations of "natural" differences, see Cynthia Eagle Russett, *Sexual Science: The Victorian Construction of Womanhood* (Cambridge: Harvard UP, 1989).

9. See Mary Poovey, *Uneven Developments: The Ideological Work of Gender in Mid-Victorian England* (Chicago: U of Chicago P, 1988).

10. Martha Somerville, ed., *Personal Recollections, from Early Life to Old Age of Mary Somerville* (Boston: Roberts, 1874), 90.

11. Barbara Timm Gates, ed., *The Journal of Emily Shore* (Charlottesville: UP of Virginia, 1991), 89, 157, 59. Gates finds in Emily Shore's Victorian autobiography a record of the adolescent Shore's self-representations as "the student self, the girl-young woman self, and the dying self" (xxii).

12. Scourse, *Victorians and Their Flowers*, 77.

13. Larry J. Schaaf, *Sun Gardens: Victorian Photograms* (New York: Aperture, 1985).

14. *Quarterly Review* 101 (1857): 6. Anne Elizabeth Baker also was interested in language and provincial customs and in later years, working on her own, issued a two-volume *Glossary of Northamptonshire Words and Phrases* (1854) that included local names for plants among many examples of colloquial usage in that county. On Anne Elizabeth Baker, see also *Gentleman's Magazine*, n.s., 11 (1861): 208; and G. C. Druce, *Flora of Northamptonshire* (Arbroath: Bunch, 1930), lxxxviii–xc. On brother-sister companionship as a Victorian ideal, see Deborah Gorham, *The Victorian Girl and the Feminine Ideal* (Bloomington: Indiana UP, 1982), 44–47.

15. Richard Mabey, *The Frampton Flora* (London: Century, 1985). The watercolors (not intended for publication) show "a comprehensive picture of the vegetation of their home countryside . . . also a fair portrait of the state of lowland England's flora in the middle of the nineteenth century" (16).

16. Davidoff and Hall, *Family Fortunes*, esp. 309–11.

17. On Maria Turner Hooker and Frances Henslow Hooker, see M. Jeanne Peterson, *Family, Love, and Work in the Lives of Victorian Gentlewomen* (Bloomington: Indiana UP, 1989), 46–47, 176.

18. Clifford B. Evans, "Stackhouse Flowers: The Life and Art of Miss Emily Stackhouse" (typescript, 1991). Mr. David Trehane of Truro, Cornwall, kindly made this material available to me.

19. Emma Peachey was scheduled to display monumental wax floral arrangements at the Great Exhibition of 1851, but the display space assigned to her was close to the roof; fearful of the effects of heat and atmospheric pollution on her specimens, she withdrew her work and displayed it afterward at her home in London. See Emma Peachey, *The Royal Guide to Wax Flower Modelling* (London, 1851), 58–68.

20. See, for example, *Curtis's Botanical Magazine*, 1826, 2699, and Charlotte Yeldham, *Women Artists in Nineteenth-Century France and England* (New York: Garland, 1984), 1:157–58.

21. Deborah Cherry, *Painting Women: Victorian Women Artists* (London: Routledge, 1993), 25. The Mutrie sisters were among women who signed a petition to members of the Royal Academy of Arts in 1855 in support of opening art schools to women.

22. "For myself I have this Spring become *moss mad*—the idea of the extreme difficulty of defining the species has always deterred me from commencing the study of this most interesting class of organized beings—having been successful in naming correctly some that grew in my own vicinity I treated myself with Hooker's Muscologia Britannica which lead me on to investigate all that came in my way—I have already obtained thirty species and displayed them on stout wire wove paper that will bear handling." Letter from Elizabeth Reynolds, July 25, 1836, cited in David Elliston Allen and Dorothy W. Lousley, "Some Letters to Margaret Stovin (1756?–1846), Botanist of Chesterfield," *Naturalist* 104 (1979): 159.

23. See David Elliston Allen, "The First Woman Pteridologist," *British Pteridological Society Bulletin* 1, 6 (1978): 247–49.

24. Horwood and Noel, *Flora of Leicestershire and Rutland*, cclxxvi.

25. See Lousley, "Some Letters to Margaret Stovin," and Mark Simmons, *A Cata-*

logue of the Herbarium of the British Flora Collected by Margaret Stovin (1756–1846) (Middlesbrough: Dorman Museum, 1993).

26. See David Elliston Allen, "The Botanical Family of Samuel Butler," *Journal of the Society for the Bibliography of Natural History* 9, 2 (1979): 134–35. She reported in a note to the *Phytologist* having visited "the Nottingham meadows for the long-desired pleasure of gathering *crocus vernus*" and inquires about the cycle of *Crocus nudiflorus* (1 [1842]: 167–68).

27. *Phycologia Britannica*, vol. 1, pl. 318.

28. Charles Kingsley, *Glaucus, or The Wonders of the Shore* (London, 1855), 54.

29. The poem, based on Charles Kingsley's "The Sands of Dee," is quoted in Christabel Maxwell, *Mrs. Gatty and Mrs. Ewing* (London: Constable, 1949), 98–99.

30. Margaret Gatty, *British Sea-Weeds: Drawn from Professor Harvey's "Phycologia Britannica"* (London, 1862), 1:ix. On Margaret Gatty, see Susan Drain, "Marine Botany in the Nineteenth Century: Margaret Gatty, the Lady Amateurs and the Professionals," *Victorian Studies Association Newsletter* (Ontario, Canada) 53 (1994): 6–11.

31. On Elizabeth Andrew Warren, see Isabella Gifford, "Memorial of Miss Warren," *Report of Royal Cornwall Polytechnic Society*, 1864, 11–14; rpt. *Journal of Botany* 3 (1865): 101–3. Also F. Hamilton Davey, *Flora of Cornwall* (London, 1909), xliii, and E[mily] S[tackhouse], "Memorial Sketch of Miss Warren of Flushing," *Journal of the Royal Institution of Cornwall*, October 1865, xviii. Letter to William Hooker dated August 22, 1837, in Hooker, *Director's Correspondence*, Royal Botanic Garden, Kew.

32. *Schizosiphon warreniae*, named by Robert Caspary in 1850, now is called *Rivularia biasolettiana*. Warren's article "Marine Algae, Found on the Falmouth Shores," appeared in *Report of the Royal Cornwall Polytechnic Society*, 1849, 31–37. In an earlier account for that publication, "Miss Warren" commented on cryptogamic plants recently found near Penzance and displayed on the natural history table at the local horticultural society. "On the recent botanical discoveries in Cornwall," 1842, 24–25.

33. Letter dated January 24, 1837. Hooker, *Director's Correspondence*, Royal Botanic Garden, Kew. There are forty-eight letters from Elizabeth Warren to William Hooker, dating December 1, 1834 to January 17, 1858.

34. Letter dated February 22, 1836, ibid.

35. Letter dated February 23, 1839, ibid.

36. *Annals and Magazine of Natural History* 3 (1839): 121–22.

37. Letter dated February 22, 1836. Hooker, *Director's Correspondence*.

38. Evans, "Stackhouse Flowers."

39. *Journal of Botany* 3 (1865): 102. In a memorial statement, the Cornish artist Emily Stackhouse recalled Elizabeth Warren's "untiring zeal in the cause of botanical science, her kindness in assisting her less informed friends, and, above all, her modest unassuming character." *Journal of the Royal Institution of Cornwall*, October 1865, xviii.

40. Letter dated December 18?, 1863, Walker Arnott Correspondence, Natural History Museum, London.

41. "Isabella Gifford," *Journal of Botany* 30 (1892): 81–83.

42. Pratt, *Imperial Eyes*.

43. On Lady Amherst, Lady Canning, and other colonial women who collected and drew the flora of the Indian subcontinent, see Desmond, *European Discovery of the Indian Flora*, 181–83 and passim. Also Augustus J. C. Hare, *The Story of Two Noble Lives: Being Memorials of Charlotte, Countess Canning, and Louisa, Marchioness of Waterford* (London: George Allen, 1893), 2:66, 3:152.

44. Mrs. James Cookson, *Flowers Drawn and Painted after Nature in India* [London, 1835].

45. *Botanical Magazine*, 2751, 2817, 2970; W. H. Harvey, "Notice of a Collection of Algae, Communicated to Dr. Hooker by the Late Mrs. Charles Telfair, from 'Cap Malheureux,' in the Mauritius; with Descriptions of Some New and Little Known Species," *Journal of Botany* 1 (1834): 147–57; letter to William Hooker dated August 7, 1829, in Hooker, *Director's Correspondence*, Royal Botanic Garden, Kew, 52:41. *Botanical Magazine*, 2976, depicts *Bignonia telfairiae* ("Mrs. Telfair's Bignonia"), a flowering plant from Madagascar named "in testimony of [Professor Bojer's] high respect for her many virtues and accomplishments, and in acknowledgement of the services rendered by her to Botany in many ways, but in none more than by her happy talent in the delineation of plants."

46. Letters to William Hooker in Hooker, *Director's Correspondence*, Royal Botanic Garden, Kew, 1:273; 2:188–91; 51:45; 53:129–33, 137; 54:528–37A. "Journal of an Ascent to the Summit of Adam's Peak, Ceylon," *Companion to the Botanical Magazine* 1 (1835): 3–14; "Journal of a Tour in Ceylon" ibid., 2 (1840): 233–56.

47. On Catharine Parr Traill, see Marianne Gosztonyi Ainley, "Science in Canada's Backwoods: Catharine Parr Traill (1802–1899)," in *Science in the Vernacular*, ed. Barbara T. Gates and Ann B. Shteir (forthcoming).

48. See *Recollections of a Happy Life, Being an Autobiography of Marianne North*, ed. Mrs. John Addington Symonds, 2 vols. (London, 1892).

Chapter 8. Flora's Daughters in Print Culture

1. On the sociology of authorship for women in the Victorian period, see Tuchman, *Edging Women Out*, and Mumm, "Writing for Their Lives."

2. Eliza Eve Gleadall, *The Beauties of Flora*, 2 vols. (Wakefield, 1834–37). A prospectus from the 1830s announces that Miss Simpson and Miss Gleadall have "a vacancy for one or two Young Ladies, who may be very eligibly accommodated as PARLOUR BOARDERS, and receive private Lessons." Along with tuition in English grammar, history, reading, and "fashionable works," they also offered French, Italian, music, drawing, dancing, writing, geography, and astronomy.

3. [Charlotte Elizabeth Tonna], *Chapters on Flowers*, 3d ed. (London, 1839), 60, 245. Other books that combine floral art and verse in "floral portraits" with the emblematic tradition are Louisa Twamley, *The Romance of Nature* (London, 1836), and her *Flora's Gems, or The Treasures of the Parterre* (London, 1837); and Rebecca Hey, *The Moral of Flowers* (London, 1833), and her *The Spirit of the Woods* (London, 1837).

4. Christine L. Krueger, *The Reader's Repentance: Women Preachers, Women Writers, and Nineteenth-Century Social Discourse* (Chicago: U of Chicago P, 1992), chap. 7.

5. Sarah Waring, *A Sketch of the Life of Linnaeus* (London: W. Darton, 1827), ix. Sarah Waring also wrote *The Wild Garland, or Prose and Verse Illustrative of English*

Wild Flowers (London, 1827; 2d ed., 1837), a book that combines romantic verse, religious content, and botanical information.

6. Louisa Ann Twamley, *Our Wild Flowers, Familiarly Described* (London, 1839), preface. Another example, Emily Ayton's *Words by the Way-Side, or The Children and the Flowers* (London, 1855), is a narrative about children and their governess, who teaches moral, religious, and Linnaean lessons.

7. Mrs. E. E. Perkins, *The Elements of Botany* (London: Thomas Hurst, 1837), preface.

8. Review of George Banks, *An Introduction to the Study of English Botany, Gardeners' Chronicle* 9 (1833): 453.

9. Perkins, *Elements of Botany*, xix.

10. Ibid., 137–38.

11. Ascribed to Perkins as well are *Elements of Drawing* (n.d.), *Haberdashery and Hosiery* (1853), and *Flora and Thalia, or Gems of Flowers and Poetry* (n.d.).

12. Cited in Morag Shiach, *Discourse on Popular Culture: Class, Gender and History in Cultural Analysis, 1730 to the Present* (Cambridge: Polity, 1989), 27.

13. Anne Pratt, *The Flowering Plants and Ferns of Great Britain*, 5 vols. (London: Society for Promoting Christian Knowledge [SPCK], 1855), 3:15; idem, *The Green Fields and Their Grasses* (London: SPCK, 1852), preface; idem, *The Ferns of Great Britain* (London: SPCK, 1855), 1.

14. Anne Pratt, *Wild Flowers* (London: Society for Promoting Christian Knowledge, 1852–53). There are nearly one hundred pages of drawings in each volume, with two pages of text about each.

15. James Britten, "Anne Pratt," *Journal of Botany* 32 (1894): 205–7; *Dictionary of National Biography*, 1921, 16:284–85; *Women's Penny Paper*, November 9, 1889; Margaret Graham, "A Life among the Flowers of Kent," *Country Life* 161 (1977): 1500.

16. Anne Pratt, *Dawnings of Genius, or The Early Lives of Some Eminent Persons of the Last Century* (1841), preface.

17. I draw here on Jonathan Topham's argument in "Science and Popular Education in the 1830s: The Role of the *Bridgewater Treatises*," *British Journal for the History of Science* 25 (1992): 397–430.

18. Anne Pratt, *Chapters on the Common Things of the Sea-Side* (London: SPCK, 1850), 1, 32, 36.

19. *Women's Penny Paper*, November 9, 1889.

20. *Journal of Botany* 32 (1894): 205–6.

21. Jackson, *Pictorial Flora*, iii–iv.

22. Elizabeth Twining, *Illustrations of the Natural Orders of Plants* (London, 1849–55), introduction.

23. Allan Bird, *Arabella Roupell: Pioneer Artist of Cape Flowers* (Johannesburg: SANH, 1975).

24. Schaaf, *Sun Gardens*, 7.

25. According to David E. Allen, "The number of species of fungi known to occur in Britain was . . . multiplied fourfold during the reign of Queen Victoria" (*Naturalist in Britain*, 128).

26. Mrs. T. J. Hussey [Anna Maria], *Illustrations of British Mycology, Containing*

Figures and Descriptions of the Funguses of Interest and Novelty Indigenous to Britain (London: Lovell Reeve, 1847–49, 1855), ser. 2 (1855), pl. 25. The work appeared in two expensive quarto volumes. Series 1 (1847–49), containing ninety plates, was published by subscription. Series 2 (1855), consisting of fifty plates, was issued in monthly numbers, each with three plates, at a price of 5s. The price for the complete work was £7 12s. 6d.

27. *Journal of Botany* 24 (1886): 252.

28. Hussey, *Illustrations of British Mycology*, ser. 1, pl. 1; ser. 2, pl. 1; ser. 1, pl. 31.

29. Ibid., ser. 1, pl. 5.

30. Gwen Raverat, *Period Piece: A Cambridge Childhood* (London: Faber and Faber, 1987), 135–36. I am grateful to Janet Browne for bringing this anecdote to my notice.

31. Letters dated April 10 and August 16, 1846, in Berkeley Correspondence, Natural History Museum, London.

32. Letter dated April 20, 1849, ibid.

33. In an article for *Hooker's London Journal of Botany* about fungi in Ceylon, when describing one new genus in great detail, Berkeley wrote: "I have named the genus after my friend Mrs. Hussey, whose talents well deserve such a distinction" (5 [1847]: 508–9). Announcing this to her by letter, he wrote: "I hope you will think it worthy of having your name. It is a strange puffball from Adam's Peak Ceylon, abounding in points of interest." Unpublished letter dated August 27, 1847; Library, Wellcome Institute for the History of Medicine, London.

34. Letters dated January 27, January 30, and May 27, 1847 in Berkeley Correspondence.

35. Thomas J. Hussey, rector of Hayes, Kent, was from an old family of Kent and Suffolk gentry that included plant collectors and a nurseryman. James Hussey, her contemporary, was a member of the Botanical Society of London (see Allen, *Botanists*). I am grateful for information about the Hussey family supplied by H. J. M. Symons, Library, Wellcome Institute for the History of Medicine, London.

36. Hussey's story "Matrimony" concerns "the chequered narrative of Matrimony," focusing on a youthful marriage made without family support, and about the tensions ensuing when "the delusion of 'love' was over." The story appeared in vols. 39–40. Themes include illegitimacy and family scheming, a fashionable family ostracizing a daughter-in-law; an older woman, "most wise and kind," contrasted with the preoccupations of fashionable wives. Her narrative features were topical during a decade of increasing debate about marriage laws, but she also admitted to Berkeley that the story had autobiographical features and that "personal feelings of the most painful kind entered into the tale" (letter dated May 1, 1849, in Berkeley Correspondence).

37. Letter dated April 10, 1846, ibid.

38. Andrew Bloxom, for example, a Leicestershire clergyman, wrote about local botany and was "perhaps the last of the all-round British naturalists" (*Dictionary of National Biography*, 1921, 2:726). Mary Kirby, acknowledging his help in her preface, wrote that she submitted "Specimens of every doubtful plant" to this "careful and experienced observer."

39. "Every Saturday night, a number of the Penny and Saturday Magazine used

to come out of our father's coat pocket. . . . and by and bye [*sic*] 'Chamber's Miscellany,' and later on, Knights publications." Mary Kirby, *"Leaflets from My Life": A Narrative Autobiography* (London: Simpkin and Marshall, 1887), 13. Mary Kirby's autobiography is a rich source for the social history of middle-class life in the Midlands during the mid-nineteenth century. See also J. D. Bennett, "Mary Kirby: A Biographical Note" (typescript, January 1965), Leicestershire Record Office Library, pamphlet box 24A.

40. Kirby, *"Leaflets from My Life,"* 40, 43, 61.

41. "Revivifying Property of the Leicestershire Udora," *Phytologist* 3 (1848): 30.

42. Kirby, *"Leaflets from My Life,"* 63, 70.

43. As professional writers for the juvenile market, Mary Kirby and Elizabeth Kirby collaborated on natural history books, moral tales, books adapted from the classics, and also stories for magazines. Their books, widely reprinted in England and America, include *Caterpillars, Butterflies, and Moths* (1857) and *Aunt Dorothy's Story Book* (1862).

44. Mary Kirby, with Elizabeth Kirby, *Plants of the Land and Water* (London: Jarrold, 1857); Kirby, *"Leaflets from My Life,"* 144.

45. Kirby, *"Leaflets from My Life,"* 165, 92.

46. Ibid., 146–48. A twentieth-century survey of the history of county flora in Leicestershire put Mary Kirby's work into perspective in this way: "Considering the time (1850) [*A Flora of Leicestershire*] was a work of definite scientific value. . . . It shows a considerable intimacy with the more critical genera, and the number of varieties included proves that Miss Kirby took pains to discriminate between typical plants and their varieties." Horwood and Noel, *Flora of Leicestershire and Rutland,* cxii–iii.

47. Kirby, *"Leaflets from My Life,"* 126.

48. On Jane Loudon, see Howe, *Lady with Green Fingers*; G. E. Fussell, "A Great Lady Botanist," *Gardeners' Chronicle* 138, 3 (1955): 192; and Geoffrey Taylor, *Some Nineteenth Century Gardeners* (London, 1951), 17–39. Jane Loudon's applications to the Royal Literary Fund supply additional information about her writing career; see *Archives of the Royal Literary Fund*, files 648 and 1101.

49. Jane Loudon's "A Short Account of the Life and Writings of John Claudius Loudon" (1845) is reprinted in John Gloag, *Mr. Loudon's England: The Life and Work of John Claudius Loudon and His Influence on Architecture and Furniture Design* (Newcastle upon Tyne: Oriel, 1970), 182–219. See also Priscilla Boniface, ed., *In Search of English Gardens: The Travels of John Claudius Loudon and His Wife Jane* (St. Albans: Lennard, 1987).

50. Jane Loudon, *British Wild Flowers* (London, 1844), 1–2.

51. Jane Loudon, *Botany for Ladies* (London: John Murray, 1842), vi.

52. "How Should Girls Be Educated?" *Ladies' Companion at Home and Abroad* 1 (1850): 184.

53. She had received assistance earlier from the Royal Literary Fund when, as Jane Webb, she applied in 1829 and was awarded twenty-five pounds. The award, she later wrote, "saved my life" (*Archives of the Royal Literary Fund*, file 648).

54. John Claudius Loudon died before he had paid off debts that arose from several works published "on his own account." Thirteen copyrights were being "held in trust

by Messrs Longmans till all his publishing debts [3,207 pounds sterling] are paid" (*Archives of the Royal Literary Fund*, file 1101, application dated May 1, 1844).

55. *Ladies' Companion* 1 (1849): 8; 1 (1850): 264.

56. *Ladies' Companion* 1 (1850): 176.

57. See Felicity Hunt, "Divided Aims: The Educational Implications of Opposing Ideologies in Girls' Secondary Schooling, 1850–1940," in *Lessons for Life: The Schooling of Girls and Women, 1850–1950*, ed. Felicity Hunt (Oxford: Blackwell, 1987).

58. "Women's Books and Men's Books," *Ladies' Companion* 1 (1850): 76.

59. Lydia Ernestine Becker, *Botany for Novices: A Short Outline of the Natural System of Classification of Plants* (London: Whittaker, 1864), iii–iv.

60. On Lydia Becker, see *Dictionary of National Biography*, 1921; and Helen Blackburn, *Women's Suffrage: A Record of the Women's Suffrage Movement in the British Isles, with Biographical Sketches of Miss Becker* (London: Williams and Norgate, 1902; rpt. New York, 1971).

61. On "local government ladies" on school boards and poor law boards, see Patricia Hollis, "Women in Council: Separate Spheres, Public Spaces," in *Equal or Different: Women's Politics, 1800–1914*, ed. Jane Rendall (Oxford: Blackwell, 1987), 192–213.

62. On women's education in England during the 1860s and 1870s, see Margaret E. Bryant, *The Unexpected Revolution: A Study in the History of the Education of Women and Girls in the Nineteenth Century* (London: University of London Institute of Education, 1979), and *The London Experience of Secondary Education* (London: Athlone, 1986), chap. 7. See also June Purvis, *Hard Lessons: The Lives and Education of Working-Class Women in Nineteenth-Century England* (Cambridge: Polity, 1989), chap. 5.

63. Lydia Ernestine Becker, "On the Study of Science by Women," *Contemporary Review* 10 (1869): 386.

64. John Leigh, physician, chemist, man of science, and after 1868, Manchester's first medical health officer, also was active in the Manchester Literary and Philosophical Society. Robert H. Kargon, *Science in Victorian Manchester: Enterprise and Expertise* (Manchester: Manchester UP, 1977), 66–74. Letters to Lydia Becker from John Leigh, in the "Becker Letters," formerly in the Fawcett Library. In addition to writing *Botany for Novices*, Lydia Becker also wrote an introductory book on astronomy that was never published.

65. Darwin sent her papers titled "Climbing Plants" and "On the Sexual Relations of the Three Forms of *Lythrum salicaria*." See Frederick Burkhardt and Sydney Smith, eds., *A Calendar of the Correspondence of Charles Darwin, 1821–1882* (New York: Garland, 1985), no. 5391. There are fourteen letters from Lydia Becker to Charles Darwin, dating between 1863 and 1877.

66. Reported in *Journal of Botany* 7 (1869): 292. Lydia Becker later wrote to Charles Darwin for advice about where to submit her paper for publication and requested permission to quote extracts from his letters to her (letter dated December 29, 1869).

67. Letter dated April 5, 1867, from Charles C. Babington; formerly in the Fawcett Library.

68. Letter to Mrs. Henry Fawcett, February 6, 1887, formerly in the Fawcett Library.

69. Blackburn, *Women's Suffrage*, 30.

Epilogue: Flora Feministica

1. *The Journal of Beatrix Potter from 1881 to 1897*, ed. Leslie Linder (London: Warne, 1966), 413–14 and 423–30. See also *Beatrix Potter's Letters*, ed. Judy Taylor (London: Warne, 1989), 37–41.

2. On Beatrix Potter's scientific work, see Eileen Jay, Mary Noble, and Anne Stevenson Hobbs, *A Victorian Naturalist: Beatrix Potter's Drawings from the Armitt Collection* (London: F. Warne, 1992).

3. Sir Henry Roscoe was professor of chemistry at Owen's College, Manchester, and coauthor of the classic text *Treatise on Chemistry* (1877–84). Roscoe had a botanical lineage: his grandfather William Roscoe was first president of the Liverpool Royal Institution, promoter of the Liverpool Botanical Garden, and author of *Monandrian Plants of the Order Scitaminae* (1824–29), and his aunt Margaret Roscoe was a botanical artist who contributed plates to that work and also published her own *Floral Illustrations of the Seasons* (1829–31).

4. Roy MacLeod and Russell Moseley, *Days of Judgement: Science, Examinations and the Organization of Knowledge in Late Victorian England* (Driffield: Nafferton, 1982).

5. Roy MacLeod and Russell Moseley, "Fathers and Daughters: Reflections on Women, Science and Victorian Cambridge," *History of Education* 8, 4 (1979): 325.

6. See A. T. Gage and W. T. Stearn, *A Bicentenary History of the Linnean Society of London* (London: Academic, 1988), 88–93; and Margot Walker, "Admission of Lady Fellows," *Linnean* (newsletter and proceedings of the Linnean Society of London) 1, 1 (1984): 9–11.

7. See Ruth Hall, *Marie Stopes: A Biography* (London: Virago, 1977); and Peter Eaton and Marilyn Warnick, *Marie Stopes: A Checklist of Her Writings* (London: Croom Helm, 1977). Marie Stopes also founded and edited the *Sportophyte: A British Journal of Botanical Humour* (1910–13).

Bibliography

Primary Sources

Place of publication is London unless otherwise shown.

Abbot, Charles. *Flora Bedfordiensis*. 1798.

Algarotti, Francesco. *Sir Isaac Newton's Philosophy Explain'd for the Use of the Ladies.* Trans. Elizabeth Carter. 1739.

Alston, Charles. *A Dissertation on Botany.* 1754.

————. *Essays and Observations, Physical and Literary.* 2d ed. Edinburgh, 1771.

Atkins, Anna Children. *Photographs of British Algae: Cyanotype Impressions.* 1843–53.

Ayton, Emily. *Words by the Way-Side, or The Children and the Flowers.* 1855.

Banks, Joseph. Dawson Turner Copies, Banks Correspondence. Botany Library, Natural History Museum, London.

Barbauld, Anna. *Evenings at Home.* 1792.

Barker, Jane. *A Patch-Work Screen for the Ladies.* 1723; rpt. New York: Garland, 1973.

Beaufort, Harriet. *Dialogues on Botany.* 1819.

Becker, Lydia Ernestine. *Botany for Novices: A Short Outline of the Natural System of Classification of Plants.* 1864.

————. *Correspondence: "The Becker Letters."* Manuscript formerly in the Fawcett Library, London.

————. "Is There Any Specific Distinction between Male and Female Intellect?" *English Woman's Review* 3 (1868): 483–91.

————. "On the Study of Science by Women." *Contemporary Review* 10 (1869): 386–404.

Bentham, George. "Notes on *Mimoseae*, with a Short Synopsis of Species." *Journal of Botany* 15 (1842): 324.

Bingley, Rev. W. *Practical Introduction to Botany.* 1817.

Blackwell, Elizabeth. *A Curious Herbal.* 2 vols. 1737, 1739.

Bref och Skrifvelser af och til Carl von Linné. Uppsala, 1916.

Brown, G. *A New Treatise on Flower Painting, or Every Lady Her Own Drawing Master.* 3d ed. 1799.

Bryan, Margaret. *A Compendious System of Astronomy.* 1797.

————. *Lectures on Natural Philosophy.* 1806.

Bryant, Charles. *Flora Diaetetica.* 1783.

Chapone, Hester. *Letters on the Improvement of the Mind.* 1773.

————. "On Conversation." *Miscellanies in Prose and Verse.* 1775.

Clare, John. *The Letters of John Clare*. Ed. Mark Storey. Oxford: Clarendon, 1985.

———. *The Natural History Prose Writings of John Clare*. Ed. Margaret Grainger. Oxford: Clarendon, 1983.

Colden, Jane. *Jane Colden—Botanic Manuscript*. New York: Chanticleer, 1963.

Coleridge, Samuel Taylor. *The Collected Letters of Samuel Taylor Coleridge*. Ed. E. L. Griggs. Oxford: Clarendon, 1971.

Cookson, Mrs. James. *Flowers Drawn and Painted after Nature in India*. 1835.

Darwin, Erasmus. *The Botanic Garden*. Parts 1 and 2. 1791; rpt. Menston, Yorkshire: Scolar, 1973.

———. *A Plan for the Conduct of Female Education in Boarding Schools*. 1797; rpt. New York: Johnson, 1968.

———. *Zoonomia; or The Laws of Organic Life*. 2d ed. 1796.

Delany, Mary. *The Autobiography and Correspondence of Mary Granville, Mrs. Delany*. Ed. Lady Llanover. Ser. 1, 1861; ser. 2, 1862.

Drummond, James. *First Steps to Botany*. 1823.

Edgeworth, Maria. *Letters for Literary Ladies*. 1795.

———. *Maria Edgeworth in France and Switzerland: Selections from the Edgeworth Family Letters*. Ed. Christina Colvin. New York: Oxford UP. 1979.

———. *Maria Edgeworth: Letters from England, 1813–1844*. Ed. Christina Colvin. Oxford: Clarendon, 1971.

———. *Practical Education*. 1798.

Ehret, George. "A Memoir of George Dionysius Ehret." 1758. *Proceedings of the Linnean Society*, 1894–95, 41–58.

Elegant Arts for Ladies. ca. 1856.

Euler, Leonhard. *Letters of Euler on Different Subjects in Natural Philosophy, Addressed to a German Princess*. Ed. David Brewster. New York, 1833.

———. *Letters to a German Princess on Different Subjects in Physics and Philosophy*. Trans. Henry Hunter. 1768.

Farquhar, George. *The Works of George Farquhar*. Ed. Shirley Strum Kenny. Oxford: Clarendon, 1988.

Fennell, James H. *Drawing-Room Botany*. 1840.

Ferguson, James. *The Young Gentleman and Lady's Astronomy*. 1768.

Fitton, Sarah. *Conversations on Botany*. 2d ed. 1818.

———. *The Four Seasons: A Short Account of the Structure of Plants*. 1865.

———. *How I Became a Governess*. 1861.

The Floral Knitting Book. 1847.

Fontenelle, Bernard le Bovier de. *Conversations on the Plurality of Worlds*. Trans. H. A. Hargreaves. Introduction by Nina R. Gelbart. Berkeley and Los Angeles: U of California P, 1990.

Forsyth, John S. *The First Lines of Botany, or Primer to the Linnaean System*. 1827.

Francis, G. *The Grammar of Botany*. 1840.

Gaskell, Elizabeth. *Cranford*. 1853.

Gatty, Margaret. *British Sea-Weeds: Drawn from Prof. Harvey's "Phycologia Britannica."* 1862.

Gerard, John. *The Herball, or Generall Historie of Plants*. 1597.

Gifford, Isabella. *The Marine Botanist.* 1848.

———. "Memorial of Miss Warren." *Report of Royal Cornwall Polytechnic Society,* 1864, 11–14; rpt. *Journal of Botany* 3 (1865): 101–3.

Gleadall, Eliza Eve. *The Beauties of Flora, with Botanic and Poetic Illustrations: Being a Selection of Flowers Drawn from Nature, Arranged Emblematically with Directions for Coloring Them.* 1834–37.

Gregory, G. *The Economy of Nature.* 1796.

Grey, Elizabeth Talbot, Countess of Kent. *Choice Manuall, or Rare and Select Secrets in Physick and Chyrurgery.* 2d ed. 1653.

Halsted, Caroline A. *The Little Botanist, or Steps to the Attainment of Botanical Knowledge.* 1835.

Hardcastle, Lucy. *An Introduction to the Elements of the Linnaean System of Botany.* 1830.

Heckle, Augustin. *The Lady's Drawing Book.* 1753.

Henfrey, Arthur. *Elementary Course of Botany.* 1857.

Hey, Rebecca. *The Moral of Flowers,* 1833.

———. *The Spirit of the Woods.* 1837; new ed. 1849, *Sylvan Musings.*

Hoare, Sarah. *A Poem on the Pleasures and Advantages of Botanical Pursuits.* 1826.

———. *Poems on Conchology and Botany.* 1831.

Hooker, Sir William J. *British Flora.* 1830; 5th rev. ed., 1842.

———. *Director's Correspondence.* Kew: Royal Botanic Gardens.

Hughes-Gibb, Eleanor. *The Making of a Daisy, "Wheat out of Lilies," and Other Studies in Plant-Life and Evolution: A Popular Introduction to Botany.* 1898.

Hunt, Leigh. *The Correspondence of Leigh Hunt.* 2 vols. 1862.

———. *Foliage, or Poems Original and Translated.* 1818.

Hussey, Mrs. T. J. [Anna Maria]. *Illustrations of British Mycology, Containing Figures and Descriptions of the Funguses of Interest and Novelty Indigenous to Britain.* 1847–49, 1855.

———. Letters to Miles Berkeley, 1843–49. Berkeley Correspondence, Natural History Museum, London.

Ibbetson, Agnes. 5 vols. of manuscripts. Natural History Museum, London (MSS IBB).

———. Essays in *Annals of Philosophy.* 1818–19. Vols. 11–14 passim.

———. Essays in William Nicholson's *Journal of Natural Philosophy, Chemistry, and the Arts.* 1809–13. Vols. 23–36 passim.

———. Essays in *Philosophical Magazine,* 1814–22. Vols. 43–60 passim.

———. "On the Adapting of Plants to the Soil, and Not the Soil to the Plants." *Letters and Papers on Agriculture, Planting, etc. Selected from the Correspondence of the Bath and West of England Society* 14 (1816): 136–59.

———. "On the Structure and Growth of Seeds." *Journal of Natural Philosophy, Chemistry, and the Arts.* September 1810.

———. "Phytology" and letters. Linnean Society, London (MSS. 120a, 120b, 490).

Jackson, Mary Anne. *The Pictorial Flora, or British Botany Delineated, in 1500 Lithographic Drawings of All Species of Flowering Plants Indigenous to Great Britain.* 1840.

Jacson, Maria Elizabeth. *Botanical Dialogues, between Hortensia and Her Four Children.* London: J. Johnson, 1797.

————. *Botanical Lectures.* 1804.

————. *A Florist's Manual: Hints for the Construction of a Gay Flower-Garden.* 1816.

————. *Sketches of the Physiology of Vegetable Life.* 1811.

Jacson, Frances. "Diaries of Frances Jacson, 1829–37." Lancashire Record Office, Preston, Lancs. DX 267–78.

Jacson, Simon. Wills dated February 1796 and April 1799. Cheshire Record Office, Chester, DDX 9/19.

Keats, John. *The Letters of John Keats.* Ed. Maurice Buxton Forman. London: Oxford UP, 1952.

Kent, Elizabeth. *Flora Domestica, or The Portable Flower-Garden.* 1823.

————. "The Florist." In *The Young Lady's Book: A Manual of Elegant Recreations, Arts, Sciences, and Accomplishments.* 1829.

————. "An Introductory View of the Linnaean System of Plants." *Magazine of Natural History* vols. 1–3 (1828–30), passim.

————, ed. *Synoptical Compendium of British Botany,* by John Galpine. 1834.

————. *Sylvan Sketches, or A Companion to the Park and Shrubbery.* 1825.

Kingsley, Charles. *Glaucus, or The Wonders of the Shore.* 1855.

————. *Madame How and Lady Why, or First Lessons in Earth Lore for Children.* 1869.

Kirby, Mary. *"Leaflets from My Life": A Narrative Autobiography.* 1887.

————. "Revivifying Property of the Leicester Udora." *Phytologist* 3 (1848): 30.

————, with Elizabeth Kirby. *Chapters on Trees: A Popular Account of Their Nature and Uses.* 1873.

————. *A Flora of Leicestershire.* 1850.

————. *Plants of the Land and Water.* 1857.

Laurie, Charlotte L. *Flowering Plants: Their Structure and Habitat.* 1903.

Lee, James. *Introduction to Botany: Extracted from the Works of Dr. Linnaeus.* 1760.

Liebig, Justus von. *Familiar Letters on Chemistry.* 1843–44.

Lindley, John. *Introduction to Botany.* 1832.

————. *An Introductory Lecture Delivered in the University of London on Thursday, April 30, 1829.* 1829.

————. *Ladies' Botany, or A Familiar Introduction to the Study of the Natural System of Botany.* 2 vols. 1834–37.

————. *School Botany.* 1839.

Linnaeus, Carolus. *A Dissertation on the Sexes of Plants.* Trans. James Edward Smith. 1786.

————. *Species Plantarum.* 1753; rpt. London: Ray Society, 1957–59.

————. *A System of Vegetables.* Trans. Botanical Society of Lichfield. 1783.

Linnea, Elisabeth Christina. "Om Indianska Krassens Blickande." *Svenska Kongliga Vetenskaps Academiens Handlingar* 23 (1762): 284–86.

Loudon, Jane. *Botany for Ladies,* 1842; 2d ed., 1851, *Modern Botany.*

————. *British Wild Flowers.* 1844.

————. *Conversations upon Chronology and General History, from the Creation of the World to the Birth of Christ.* 1830.

———. *First Book of Botany.* 1841.

———. "A Short Account of the Life and Writings of John Claudius Loudon." 1845. Rpt. in John Gloag, *Mr. Loudon's England: The Life and Work of John Claudius Loudon and His Influence on Architecture and Furniture Design.* Newcastle upon Tyne: Oriel Press, 1970.

Mangnall, Richmal. *Historical and Miscellaneous Questions for the Use of Young People.* 1800.

Marcet, Jane. *Conversations on Chemistry.* 1806.

———. *Conversations on Natural Philosophy.* 1819.

———. *Conversations on Political Economy.* 1816.

———. *Conversations on Vegetable Physiology.* 1829.

———, trans. *General Observations on Vegetation,* by C. F. Brisseau de Mirbel. 1833.

Martyn, Thomas. *Letters on the Elements of Botany.* 1785.

Mavor, William. *A Catechism of Botany . . . for the Use of Schools and Families.* 1800.

———. *The Lady's and Gentleman's Botanical Pocket Book.* 1800.

Mitford, Mary Russell. *Poems.* 1810.

———. *The Works of Mary R. Mitford.* 1850.

More, Hannah. *Strictures on the Modern System of Female Education.* 1799. In *The Works of Hannah More,* vol. 3. 1853.

Moriarty, Henrietta Maria. *Brighton in an Uproar.* 1811.

———. *Crim. Con. A Novel, Founded on Facts.* 1812.

———. *A Hero of Salamanca, or The Novice Isabel.* 1813.

———. *Viridarium.* 1806; 2d ed., *Fifty Green-House Plants.* 1807.

Murray, Lady Charlotte. *The British Garden.* 1799.

Murry, Ann. *Mentoria, or The Young Ladies Instructor.* 1778.

———. *Sequel to Mentoria.* 1799.

Newbery, John. *The Newtonian System of Philosophy Adapted to the Capacities of Young Gentlemen and Ladies.* 1761.

North, Marianne. *Recollections of a Happy Life, Being an Autobiography of Marianne North.* Ed. Mrs. John Addington Symonds. 2 vols. 1892.

Oliver, Daniel. *Lessons in Elementary Botany: The Part of Systematic Botany Based upon Material Left in Manuscript by the Late Professor Henslow.* 1864.

Parkes, Samuel. *The Chemical Catechism for the Use of Young People.* 1808.

Peachey, Emma. *The Royal Guide to Wax Flower Modelling.* 1851.

Perkins, Mrs. E. E. *The Elements of Botany.* 1837.

Perry, James. *Mimosa, or The Sensitive Plant.* 1779.

Polwhele, Richard. *The Unsex'd Females.* 1798.

Potter, Beatrix. *Beatrix Potter's Letters.* Ed. Judy Taylor. London: Warne, 1989.

———. *The Journal of Beatrix Potter from 1881 to 1897.* Ed. Leslie Linder. London: Warne, 1966.

Pratt, Anne. *Chapters on Common Things of the Sea-Side.* 1850.

———. *The Ferns of Great Britain, and Their Allies the Club Mosses, Pepperworts and Horsetails.* 1855.

———. *The Field, the Garden, and the Woodland, or Interesting Facts Respecting Flowers and Plants in General.* 1838.

————. *The Flowering Plants and Ferns of Great Britain.* 5 vols. 1855.

————. *Flowers and Their Associations.* 1840.

————. *The Green Fields and Their Grasses.* 1852.

————. *Poisonous, Noxious, and Suspected Plants of Our Fields and Woods.* 1857.

————. *Wild Flowers.* 1852–53.

Pulteney, Richard. *Historical and Biographical Sketches of the Progress of Botany in England from Its Origin to the Introduction of the Linnaean System.* 1790.

Ralph, T. S. *Elementary Botany.* 1849.

Roberts, Mary. *Annals of My Village: Being a Calendar of Nature.* 1831.

————. *Flowers of the Matin and Even Song, or Thoughts for Those Who Rise Early.* 1845.

————. *Select Female Biography, Comprising Memoirs of Eminent British Ladies.* 1821.

————. *Sister Mary's Tales in Natural History.* 1834.

————. *Wonders of the Vegetable Kingdom Displayed.* 1822.

Rootsey, S. *Syllabus of a Course of Botanical Lectures.* Bristol, 1818.

Roupell, Arabella. *Specimens of the Flora of South Africa by a Lady.* 1849.

Rousseau, Jean-Jacques. *Lettres élémentaires sur la botanique.* 1771.

Rowden, Frances Arabella. *A Poetical Introduction to the Study of Botany.* 1801.

Selwyn, Amelia. *A Key or Familiar Introduction to the Study of Botany.* 1824.

Seward, Anna. *Letters Written between the Years 1784 and 1807.* Edinburgh, 1811.

————. *Memoirs of the Life of Dr. Darwin.* 1804.

Shelley, Mary Wollstonecraft. *The Letters of Mary Wollstonecraft Shelley.* Ed. Betty T. Bennett. Baltimore: Johns Hopkins UP, 1980.

Shelley, Percy Bysshe. *The Letters of Percy Bysshe Shelley.* Ed. Frederick L. Jones. Oxford: Clarendon, 1964.

Smith, Charlotte. *Conversations Introducing Poetry: Chiefly on Subjects of Natural History.* 1804.

————. *The Poems of Charlotte Smith.* Ed. Stuart Curran. New York: Oxford UP, 1993.

————. *Rural Walks.* 1795.

————. *The Young Philosopher.* 1798.

Smith, James E. *English Botany, or Coloured Figures of British Plants.* Figures by James Sowerby. 36 vols. 1790–1814.

————. *The English Flora.* 4 vols. 1824–28.

————. *An Introduction to Physiological and Systematical Botany.* 1807.

Smith, Lady Pleasance, ed. *Memoir and Correspondence of the Late Sir James Edward Smith, M.D.* 1832.

Somerville, Martha, ed. *Personal Recollections from Early Life to Old Age of Mary Somerville.* Boston, 1874.

Spence, William, and William Kirby. *Introduction to Entomology.* 1800.

S[tackhouse], E[mily]. "Memorial Sketch of Miss Warren of Flushing," *Journal of the Royal Institution of Cornwall,* October 1865, xviii.

Stillingfleet, Benjamin. *Miscellaneous Tracts relating to Natural History, Husbandry, and Physick.* 1759.

Stopes, Marie C. *Ancient Plants, Being a Simple Account of the Past Vegetation of* the Earth and of the Recent Important Discoveries Made in This Realm of Nature Study. 1910.

———. The Study of Plant Life for Young People. 1906.

Thornton, Robert John. New Illustration of the Sexual System of Carolus von Linnaeus. 1799.

Tonna, Charlotte Elizabeth. Chapters on Flowers. 1836.

Trimmer, Sarah. An Easy Introduction to the Study of Nature, and to Reading Holy Scriptures. 1780.

Twamley, Louisa Ann. Flora's Gems, or The Treasures of the Parterre. 1837.

———. Our Wild Flowers, Familiarly Described. 1839.

———. The Romance of Nature. 1836.

Twining, Elizabeth. Illustrations of the Natural Orders of Plants. 1849–55.

———. Short Lectures on Plants for Schools and Adult Classes. 1858.

Wakefield, Priscilla. Domestic Recreation, or Dialogues Illustrative of Natural and Scientific Subjects. 1805.

———. An Introduction to Botany, in a Series of Familiar Letters. 1796.

———. The Juvenile Travellers: Containing the Remarks of a Family during a Tour through the Principal States and Kingdoms of Europe. 1801.

———. Mental Improvement, or The Beauties and Wonders of Nature and Art. 1794–97; Ed. Ann B. Shteir. East Lansing: Colleagues, 1995.

Waring, Sarah. The Meadow Queen, or The Young Botanists. 1836.

———. A Sketch of the Life of Linnaeus. 1827.

———. The Wild Garland, or Prose and Verse Illustrative of English Wild Flowers. 1827.

Warren, Elizabeth Andrew. A Botanical Chart for Schools. Ca. 1830.

———. Forty-eight Letters to William Hooker. In William J. Hooker, Director's Correspondence. Kew: Royal Botanic Gardens.

———. "Hortus Siccus of the Indigenous Plants of Cornwall." 3 vols. Royal Institution of Cornwall.

———. "Marine Algae, Found on the Falmouth Shores." Report of the Royal Cornwall Polytechnic Society, 1849, 31–37.

———. "On the Recent Botanical Discoveries in Cornwall." Report of the Royal Cornwall Polytechnic Society, 1842, 24–25.

White, Gilbert. The Natural History and Antiquities of Selborne. 1789.

Williams, Anna. Miscellanies in Prose and Verse. 1766.

Wilson, Sarah Atkins. Botanical Rambles, Designed as an Early and Familiar Introduction to the Elegant and Pleasing Study of Botany. 1822.

———. A Visit to Grove Cottage. 1823.

Withering, William. Account of the Foxglove and Some of Its Medical Uses. 1785.

———. A Botanical Arrangement of All the Vegetables Naturally Growing in Great Britain. 2d ed. Birmingham, 1787.

———. Miscellaneous Tracts, to Which Is Prefixed a Memoir of His Life, Character, and Writings. Ed. William Withering the Younger. 2 vols. 1822.

Wollstonecraft, Mary. Original Stories from Real Life. 1788.

———. A Vindication of the Rights of Woman. 1792; London: Penguin Books, 1992.

———, trans. Elements of Morality, for the Use of Children, by C. G. Salzmann. 1791.

Woolley, Hannah. *Gentlewoman's Companion*. 1675.

The Young Lady's Book: A Manual of Elegant Recreations, Arts, Sciences, and Accomplishments. 1829; 2d ed., 1859.

The Young Lady's Book of Botany. 1838.

The Young Lady's Introduction to Natural History. 1766.

Periodicals and Newspapers

Annals of Philosophy
Anti-Jacobin Review
Le Beau Monde
Botanical Magazine
Botanist
British Critic
Chambers's Journal of Popular Literature
Critical Review
Eclectic Review
Examiner
Female Spectator
Gardeners' Chronicle
Gardener's Magazine
Gentleman's Magazine
Journal of Botany
Journal of Natural Philosophy, Chemistry, and the Arts
Ladies' Companion at Home and Abroad

Lady's Monthly Museum
Lady's Museum
Lady's Poetical Magazine
Literary Gazette
Magazine of Natural History
Monthly Magazine
Monthly Review
New British Lady's Magazine
New Lady's Magazine
New Monthly Magazine
Philosophical Magazine
Phytologist
Quarterly Review
Women's Penny Paper
Young Lady's Magazine of Theology, History, Philosophy and General Knowledge

Secondary Sources

Abir-Am, Pnina, and Dorinda Outram, eds. *Uneasy Careers and Intimate Lives: Women in Science, 1789–1979*. New Brunswick: Rutgers UP, 1987.

Ainley, Marianne Gosztonyi. "Science in Canada's Backwoods: Catharine Parr Traill (1802–1899)." In *Science in the Vernacular*, ed. Barbara T. Gates and Ann B. Shteir. Forthcoming.

Alic, Margaret. *Hypatia's Heritage*. London: Women's Press, 1986.

Allen, David Elliston. "The Botanical Family of Samuel Butler." *Journal of the Society for the Bibliography of Natural History* 9, 2 (1979): 133–36.

———. *The Botanists: A History of the Botanical Society of the British Isles through 150 Years*. Winchester: St. Paul's Bibliographies, 1986.

"The First Woman Pteridologist." *British Pteridological Society Bulletin* 1, 6 (1978): 247–49.

———. "The Natural History Society in Britain through the Years." *Archives of Natural History* 14 (1987): 243–59.

———. *The Naturalist in Britain: A Social History*. 1976; 2d ed., Princeton: Princeton UP, 1994.

———. *The Victorian Fern Craze: A History of Pteridomania*. London: Hutchinson, 1969.

———. "The Women Members of the Botanical Society of London, 1836–56." *British Journal for the History of Science* 13, 45 (1980): 240–54.

Allen, David Elliston, and Dorothy W. Lousley. "Some Letters to Margaret Stovin (1756?–1846), Botanist of Chesterfield." *Naturalist* 104 (1979): 155–63.

Amies, Marion. "Amusing and Instructive Conversations: The Literary Genre and Its Relevance to Home Education." *History of Education* 14, 2 (1985): 87–99.

Archives of the Royal Literary Fund, 1790–1918. London: World Microfilm, 1982.

Armstrong, Nancy. *Desire and Domestic Fiction: A Political History of the Novel*. New York: Oxford UP, 1987.

Armstrong, Nancy, and Leonard Tennenhouse, eds. *The Ideology of Conduct: Essays in Literature and the History of Sexuality*. New York: Methuen, 1987.

Bailey, Peter. *Leisure and Class in Victorian England*. London: Methuen, 1987.

Barker-Benfield, G. J. *The Culture of Sensibility: Sex and Society in Eighteenth-Century Britain*. Chicago: U of Chicago P, 1992.

Bazerman, Charles. *Shaping Written Knowledge: The Genre and Activity of the Experimental Article in Science*. Madison: U of Wisconsin P, 1988.

Benjamin, Marina, ed. *A Question of Identity: Women, Science, and Literature*. New Brunswick: Rutgers UP, 1993.

———. *Science and Sensibility: Gender and Scientific Enquiry, 1780–1945*. Oxford: Blackwell, 1991.

Bennett, J. D. "Mary Kirby. A Biographical Note." Typescript, January 1965. Leicestershire Record Office Library. Pamphlet box 24A.

Bewell, Alan. "Keats's 'Realm of Flora.'" *Studies in Romanticism* 31, 1 (1992): 71–98.

Blackburn, Helen. *Women's Suffrage: A Record of the Women's Suffrage Movement in the British Isles, with Biographical Sketches of Miss Becker*. London: Williams and Norgate, 1902; rpt. New York, 1971.

Blackwood, John. "The Wordsworths' Book of Botany." *Country Life*, October 27, 1983, 1172–73.

Blunt, Wilfrid. *The Art of Botanical Illustration*. London: Collins, 1950.

———. *The Compleat Naturalist: A Life of Linnaeus*. London: Collins, 1971.

———. *In for a Penny: A Prospect of Kew Gardens, Their Flora, Fauna and Falballas*. London: Hamish Hamilton, 1978.

Blunt, Wilfrid, and Sandra Raphael. *The Illustrated Herbal*. London: Lincoln, 1979.

Boniface, Priscilla, ed. *In Search of English Gardens: The Travels of John Claudius Loudon and His Wife Jane*. St. Albans: Lennard, 1987.

Bonta, Marcia Myers. *Women in the Field: America's Pioneering Women Naturalists*. College Station: Texas A&M UP, 1991.

Bracegirdle, Brian. *A History of Microtechnique*. Ithaca: Cornell UP, 1978.

Bradbury, S., and G. L'E. Turner, eds. *Historical Aspects of Microscopy*. Cambridge: Heffner, 1967.

Bremner, Jean P. "Some Aspects of Botany Teaching in English Schools in the Second Half of the Nineteenth Century." *School Science Review* 38 (1956–57): 376–83.

Britten, James. "Anne Pratt." *Journal of Botany* 32 (1894): 205–7.

———. "Jane Colden and the Flora of New York." *Journal of Botany* 33 (1895): 13.

———. "Lady Anne Monson." *Journal of Botany* 56 (1918): 147–49.

————. "Mrs. Moriarty's 'Viridarium.'" *Journal of Botany* 55 (1917): 52–54.

Broberg, Gunnar. "Fruntimmersbotaniken" (Botany for women). *Svenska Linnésäll-skapets Årsskrift* 12 (1990–91): 177–231.

Brody, Judit. "The Pen Is Mightier Than the Test Tube." *New Scientist*, February 14 1985, 56–58.

Browne, Janet. "Botany for Gentlemen: Erasmus Darwin and the Loves of the Plants." *Isis* 80, 4 (1989): 593–620.

Bryant, Margaret E. *The London Experience of Secondary Education*. London: Athlone, 1986.

————. *The Unexpected Revolution: A Study in the History of the Education of Women and Girls in the Nineteenth Century*. London: University of London Institute of Education, 1979.

Burke, Peter. *The Art of Conversation*. Ithaca: Cornell UP, 1993.

Burkhardt, Frederick, and Sydney Smith, eds. *A Calendar of the Correspondence of Charles Darwin, 1821–1882*. New York: Garland, 1985.

Caine, Barbara. *Victorian Feminists*. New York: Oxford UP, 1992.

Calmann, Gerta. *Ehret, Flower Painter Extraordinary: An Illustrated Biography*. Oxford: Phaidon, 1977.

Cameron, Kenneth Neill. *Shelley and His Circle, 1773–1822*. New York: Pforzheimer Library, 1961.

Candolle, Alphonse de, ed. *Mémoires et souvenirs de Augustin-Pyramus de Candolle*. Geneva, 1862.

Cantor, Geoffrey. *Michael Faraday: Sandemanian and Scientist*. London: Macmillan, 1991.

Caroe, Gwendy. *The Royal Institution: An Informal History*. London: Murray, 1985.

Carter, Harold B. *Sir Joseph Banks, 1743–1820*. London: British Museum (Natural History), 1988.

Cherry, Deborah. *Painting Women: Victorian Women Artists*. London: Routledge, 1993.

Christie, John, and Sally Shuttleworth, eds. *Nature Transfigured: Science and Literature, 1700–1900*. Manchester: Manchester UP, 1989.

Citron, Marsha J. "Women and the Lied, 1775–1850." In *Women Making Music: The Western Art Tradition*, ed. Jane Bowers and Judith Tick. Urbana: U of Illinois P, 1986.

"Civil List Pensions." *The Nineteenth Century and After* 121 (1937): 273–339.

Coombe, D. E. "The Wordsworths and Botany." *Notes and Queries* 197 (1952): 298–99.

Corson, Richard. *Fashions in Hair: The First Five Thousand Years*. London: Owen, 1965.

Cox, Virginia. *The Renaissance Dialogue: Literary Dialogue in Its Social and Political Contexts, Castiglione to Galileo*. Cambridge: Cambridge UP, 1992.

Crawford, Patricia. "Women's Published Writings, 1600–1700." In *Women in English Society, 1500–1800*, ed. Mary Prior. London: Methuen, 1985.

Cross, Nigel. *The Common Writer: Life in Nineteenth-Century Grub Street*. Cambridge: Cambridge UP, 1985.

———. *The Royal Literary Fund, 1790–1918: An Introduction to the Fund's History and Archives, with an Index of Applicants.* London: World Microfilm, 1984.

Cruickshank, Dan, and Neil Burton. *Life in the Georgian City.* London: Viking, 1990.

Cunningham, Andrew, and Nicholas Jardine, eds. *Romanticism and the Sciences.* Cambridge: Cambridge UP, 1990.

Dallmen, A. A., and W. A. Lee. "An Old Cheshire Herbarium." *Lancashire and Cheshire Naturalist* 10 (1917): 167.

Darton, F. J. H. *Children's Books in England: Five Centuries of Social Life.* 3d ed. Cambridge: Cambridge UP, 1982.

Davey, F. Hamilton. *Flora of Cornwall.* 1909.

David, Linda. *Children's Books Published by W. Darton and His Sons.* Bloomington: Lilly Library, Indiana U, 1992.

Davidoff, Leonore, and Catherine Hall. *Family Fortunes: Men and Women of the English Middle Class, 1780–1850.* Chicago: U of Chicago P, 1987.

de Almeida, Hermione. *Romantic Medicine and John Keats.* New York: Oxford UP, 1991.

Desmond, Adrian. *The Politics of Evolution: Morphology, Medicine, and Reform in Radical London.* Chicago: U of Chicago P, 1989.

Desmond, Ray. *A Celebration of Flowers: Two Hundred Years of Curtis's Botanical Magazine.* Kew: Royal Botanic Gardens, 1987.

———. *Dictionary of British and Irish Botanists and Horticulturalists, Including Plant Collectors, Flower Painters and Garden Designers.* 2d ed., London: Taylor and Francis. 1994.

———. *The European Discovery of the Indian Flora.* Kew: Royal Botanic Garden; Oxford: Oxford UP, 1992.

———. "Victorian Gardening Magazines." *Garden History* 5, 3 (1977): 47–66.

Dony, J. G. "Bedfordshire Naturalists: Charles Abbot (1761–1817)." *Bedfordshire Naturalist* 2 (1948): 38–42.

———. *Flora of Bedfordshire.* Luton, 1953.

Douglas, Aileen. "Popular Science and the Representation of Women: Fontenelle and After." *Eighteenth Century Life* 18 (May 1994): 1–14.

Drain, Susan. "Marine Botany in the Nineteenth Century: Margaret Gatty, the Lady Amateurs and the Professionals." *The Victorian Studies Association Newsletter* (Ontario, Canada) 53 (1994): 6–11.

Druce, G. C. *Flora of Bedfordshire.* 1897.

———. *Flora of Buckinghamshire.* Arbroath, 1926.

———. *Flora of Northamptonshire.* Arbroath, 1930.

Eaton, Peter, and Marilyn Warnick. *Marie Stopes: A Checklist of Her Writings.* London: Croom Helm, 1977.

Ehret, George Dionysus. "A Memoir of George Dionysus Ehret." *Proceedings of the Linnean Society of London,* 1890–1900, 41–48.

Ellison, C. C. *The Hopeful Traveller: The Life and Times of David Augustus Beaufort LL.D., 1739–1821.* Kilkenny: Boethius, 1987.

Evans, Clifford B. "Stackhouse Flowers: The Life and Art of Miss Emily Stackhouse." Typescript, [1991].

Ezell, Margaret J. M. *Writing Women's Literary History*. Baltimore: Johns Hopkins UP, 1993.

Farley, John. *Gametes and Spores: Ideas about Sexual Reproduction, 1750–1914*. Baltimore: Johns Hopkins UP, 1982.

Favret, Mary A. *Romantic Correspondence: Women, Politics and the Fiction of Letters*. Cambridge: Cambridge UP, 1993.

Fergus, Jan. *Jane Austen: A Literary Life*. London: Macmillan, 1991.

Fergus, Jan, and Janice Farrar Thaddeus. "Women, Publishers, and Money, 1790–1820." In *Studies in Eighteenth-Century Culture*, ed. John Yolton and Leslie Ellen Brown, 17:191–207. East Lansing, Mich.: Colleagues Press, 1987.

Ferguson, Allan. *Natural Philosophy through the Eighteenth Century and Allied Topics*. London: Taylor and Francis, 1972.

Ferguson, Moira. "'The Cause of My Sex': Mary Scott and the Female Literary Tradition." *Huntington Library Quarterly* 50 (1987): 359–77.

———. *Subject to Others: British Women Writers and Colonial Slavery, 1670–1834*. New York: Routledge, 1992.

Fissell, Mary. *Patients, Power, and the Poor in Eighteenth-Century Bristol*. Cambridge: Cambridge UP, 1991.

Foote, George A. "Sir Humphry Davy and His Audience at the Royal Institution." *Isis* 43 (1952): 6–12.

Ford, Brian J. *Images of Science: A History of Scientific Illustration*. London: British Library, 1992.

———. *Single Lens: The Story of the Image Produced by the Compound Microscope*. New York: Harper and Row, 1985.

Fox, Celina, ed. *London, World City, 1800–1840*. New Haven: Yale UP, 1992.

Fox, R. Hingston. *Dr. John Fothergill and His Friends*. London: Macmillan, 1919.

Frängsmyr, Tore, ed. *Linnaeus: The Man and His Work*. Berkeley and Los Angeles: U of California P, 1983.

Freeman, John. *Life of the Rev. William Kirby, M.A., Rector of Barham*. London: Longman, 1852.

Freeman, R. B. *British Natural History Books, 1495–1900: A Handlist*. Hamden, Conn.: Archon, 1980.

———. "Children's Natural History Books before Queen Victoria." *History of Education Society Bulletin* 17 (1976): 7–21; 18 (1976): 6–34.

Friendly, Alfred. *Beaufort of the Admiralty: The Life of Sir Francis Beaufort, 1774–1857*. London: Hutchinson, 1977.

Fussell, G. E. "A Great Lady Botanist [Jane Loudon]." *Gardeners' Chronicle* 138 (1955): 192.

———. "Mrs. Maria Elizabeth Jacson." *Gardeners' Chronicle* 130 (1951): 63–64.

———. "The Rt. Hon. Lady Charlotte Murray." *Gardeners' Chronicle* 128 (1950): 238–39.

Gage, A. T., and W. T. Stearn. *A Bicentenary History of the Linnean Society of London*. London: Academic, 1988.

Gardener, William. "John Lindley." *Gardeners' Chronicle* 158 (Oct. 23–Nov. 27, 1965): 386 ff.

Gascoigne, John. *Joseph Banks and the English Enlightenment: Useful Knowledge and Polite Culture.* Cambridge: Cambridge UP, 1994.

Gates, Barbara T. "Retelling the Story of Science." In *Victorian Literature and Culture,* ed. John Maynard and Adrienne Auslander Munich, vol. 21. New York: AMS, 1993.

———, ed. *The Journal of Emily Shore.* Charlottesville: UP of Virginia, 1991.

Gates, Barbara T., and Ann B. Shteir, eds. *Science in the Vernacular.* Forthcoming.

Gelpi, Barbara Charlesworth. *Shelley's Goddess: Maternity, Language, Subjectivity.* New York: Oxford UP, 1992.

Goellnicht, Donald C. *The Poet-Physician: Keats and Medical Science.* Pittsburgh: U of Pittsburgh P, 1984.

Goody, Jack. *The Culture of Flowers.* Cambridge: Cambridge UP, 1993.

Gorham, Deborah. *The Victorian Girl and the Feminine Ideal.* Bloomington: Indiana UP, 1982.

Gould, Stephen Jay. "The Invisible Woman." *Natural History* 102 (June 1993): 14–23.

Graham, Margaret. "A Life among the Flowers of Kent." *Country Life* 161 (1977): 1500–1501.

Greene, Edward Lee. *Landmarks of Botanical History.* Ed. Frank N. Egerton. 1909; rpt. Stanford: Stanford UP, 1983.

Hall, Ruth. *Marie Stopes: A Biography.* London: Virago, 1977.

Hare, Augustus J. C. *The Story of Two Noble Lives: Being Memorials of Charlotte, Countess Canning, and Louisa, Marchioness of Waterford.* London: George Allen, 1893.

Harless, Christian Friedrich. *Die Verdienste der Frauen um Naturwissenschaft und Heilkunde.* Göttingen, 1830.

Harris, Ann Sutherland, and Linda Nochlin. *Women Artists, 1550–1950.* New York: Knopf, 1977.

Harris, Barbara, and JoAnn K. McNamara, eds. *Women and the Structure of Society.* Durham: Duke UP, 1984.

Harvey-Gibson, R. J. *Outlines of the History of Botany.* London: Black, 1919.

Hawks, Ellison. *Pioneers of Plant Study.* London: Sheldon, 1928.

Hayden, Ruth. *Mrs. Delany and Her Flower Collages.* London: British Museum Press, 1992.

Hedley, Owen. *Queen Charlotte.* London: Murray, 1975.

Henrey, Blanche. *British Botanical and Horticultural Literature before 1800, Comprising a History and Bibliography of Botanical and Horticultural Books Printed in England, Scotland and Ireland from the Earliest Times until 1800.* 3 vols. London: Oxford UP, 1975.

Hilbish, Florence M. A. "Charlotte Smith, Poet and Novelist (1749-1806)." Ph.D. diss., University of Pennsylvania, 1941.

Hill, Bridget. *Women, Work, and Sexual Politics in Eighteenth-Century England.* Oxford: Blackwell, 1989.

Hilton, Boyd. *The Age of Atonement: The Influence of Evangelicalism on Social and Economic Thought, 1795–1865.* Oxford: Clarendon, 1988.

Hoare, Michael E. *The Tactless Philosopher: Johann Reinhold Forster, 1729–1798*. Melbourne: Hawthorn, 1976.

Hobby, Elaine. *Virtue of Necessity: English Women's Writing, 1649–1688*. London: Virago, 1988.

Horwood, A. R., and C. W. F. Noel. *The Flora of Leicestershire and Rutland*. London: Oxford UP, 1933.

Howe, Bea. *Lady with Green Fingers: The Life of Jane Loudon*. London: Country Life, 1961.

Hunt, Felicity, ed. *Lessons for Life: The Schooling of Girls and Women, 1850–1950*. Oxford: Blackwell, 1987.

Jackson, B. D. *George Bentham*. London: Dent, 1906.

———. *Linnaeus (afterwards Carl von Linné): The Story of His Life, Adapted from the Swedish of Theodor Magnus Fries*. London: Witherby, 1923.

Jay, Eileen, Mary Noble, and Anne Stevenson Hobbs. *A Victorian Naturalist: Beatrix Potter's Drawings from the Armitt Collection*. London: F. Warne, 1992.

Jordanova, Ludmilla. "Gender and the Historiography of Science." *British Journal for the History of Science* 26 (1993): 469–83.

———. *Sexual Visions: Images of Gender in Science and Medicine*. Madison: U of Wisconsin P, 1989.

Kargon, Robert H. *Science in Victorian Manchester: Enterprise and Expertise*. Manchester: Manchester UP, 1977.

Keener, Frederick M., and Susan E. Lorsch, eds. *Eighteenth-Century Women and the Arts*. New York: Greenwood, 1988.

Keeney, Elizabeth B. *The Botanizers: Amateur Scientists in Nineteenth-Century America*. Chapel Hill: U of North Carolina P, 1992.

Keller, Evelyn Fox. *A Feeling for the Organism: The Life and Work of Barbara McClintock*. New York: Freeman, 1983.

———. *Reflections on Gender and Science*. New Haven: Yale UP, 1985.

Kemble, Frances. *Record of a Girlhood*. London, 1878.

King-Hele, Desmond. *Doctor of Revolution: The Life and Genius of Erasmus Darwin*. London: Faber and Faber, 1977.

———, ed. *The Letters of Erasmus Darwin*. Cambridge: Cambridge UP, 1981.

Knight, Charles. *Passages of a Working Life during Half a Century*. London, 1864.

Koerner, Lisbet. "Women and Utility in Enlightenment Science." *Configurations* 3, 2 (1995): 233–55.

Kowaleski-Wallace, Elizabeth. *Their Fathers' Daughters: Hannah More, Maria Edgworth and Patriarchal Complicity*. New York: Oxford UP, 1991.

Kramnick, Isaac. "Children's Literature and Bourgeois Ideology: Observations on Culture and Industrial Capitalism in the Later Eighteenth Century." *Studies in Eighteenth-Century Culture* 12 (1983): 11–44.

Kronick, David A. *A History of Scientific and Technical Periodicals: The Origins and Development of the Scientific and Technical Press, 1665–1790*. 2d ed. Metuchen, N.J.: Scarecrow, 1976.

Krueger, Christine L. *The Reader's Repentance: Women Preachers, Women Writers, and Nineteenth-Century Social Discourse*. Chicago: U of Chicago P, 1992.

Lane, Margaret. *The Tale of Beatrix Potter.* London: Warne, 1968.

Langford, Paul. *A Polite and Commercial People: England, 1727–1783.* Oxford: Oxford UP, 1992.

Laqueur, Thomas. *Making Sex: Body and Gender from the Greeks to Freud.* Cambridge: Harvard UP, 1990.

Layton, David. *Science for the People: The Origins of the School Science Curriculum in England.* New York: Science History Publications, 1973.

Leppert, Richard. *Music and Image: Domesticity, Ideology and Socio-cultural Formation in Eighteenth-Century England.* Cambridge: Cambridge UP, 1988.

Levere, Trevor H. *Poetry Realized in Nature: Samuel Taylor Coleridge and Early Nineteenth-Century Science.* Cambridge: Cambridge UP, 1981.

Levine, Philippa. *The Amateur and the Professional: Antiquarians, Historians and Archaeologists in Victorian England, 1838–1886.* Cambridge: Cambridge UP, 1986.

Levy, Anita. *Other Women: The Writing of Class, Race, and Gender, 1832–1898.* Princeton: Princeton UP, 1991.

Lieb, Laurie. "'Amusements of No Real Estimation': Mary Delany's 'Flowers from Nature.'" Paper presented at annual meeting of the American Society for Eighteenth-Century Studies, Pittsburgh, 1991.

Lindee, M. Susan. "The American Career of Jane Marcet's *Conversations on Chemistry*, 1806–1853." *Isis* 82 (1991): 8–23.

Locke, David. *Science as Writing.* New Haven: Yale UP, 1992.

Mabberly, D. J. *Jupiter Botanicus: Robert Brown of the British Museum.* London: British Museum (Natural History), 1985.

Mabey, Richard. *The Flowering of Kew: 350 Years of Flower Paintings from the Royal Botanic Gardens.* London: Century Hutchinson, 1988.

———. *The Frampton Flora.* London: Century, 1985.

———. *A Victorian Flora: Selected from the Unpublished Flora of Caroline May.* Woodstock, N.Y.: Overlook, 1991.

MacLean, Virginia. *A Short-Title Catalogue of Household and Cookery Books Published in the English Tongue, 1701–1800.* London: Prospect, 1981.

MacLeod, Roy, and Russell Moseley. *Days of Judgment: Science, Examinations and the Organization of Knowledge in Late Victorian England.* Driffield: Nafferton, 1982.

———. "Fathers and Daughters: Reflections on Women, Science and Victorian Cambridge." *History of Education* 8, 4 (1979): 321–33.

Manthorpe, Catherine. "Reflections on the Scientific Education of Girls." *School Science Review,* March 1987, 422–31.

Martineau, Harriet. *Biographical Sketches.* New York: Hurst, 1868.

Maxwell, Christabel. *Mrs. Gatty and Mrs. Ewing.* London: Constable, 1949.

McClellan, James E. *Science Reorganized: Scientific Societies in the Eighteenth Century.* New York: Columbia UP, 1985.

McKendrick, Neil, John Brewer, and J. H. Plumb. *The Birth of a Consumer Society: The Commercialization of Eighteenth-Century England.* London: Europa, 1982.

McNeil, Maureen. *Under the Banner of Science: Erasmus Darwin and His Age.* Manchester: Manchester UP, 1987.

Mellor, Anne K., ed. *Romanticism and Feminism.* Bloomington: Indiana UP, 1988.

———. *Romanticism and Gender*. New York: Routledge, 1993.

Messer-Davidow, Ellen, David R. Shumway, and David J. Sylvan, eds. *Knowledges: Historical and Critical Studies in Disciplinarity*. Charlottesville: UP of Virginia, 1993.

Meyer, Gerald Dennis. *The Scientific Lady in England, 1650–1760: An Account of Her Rise, with Emphasis on the Major Roles of the Telescope and Microscope*. Berkeley and Los Angeles: U of California P, 1955.

Mills, Sara. *Discourses of Difference: An Analysis of Women's Travel Writing and Colonialism*. London: Routledge, 1991.

Mitchell, M. E. "The Authorship of *Dialogues on Botany*." *Irish Naturalists' Journal* 19, 11 (1979): 407.

Montluzin, Emily L. de. *The Anti-Jacobins, 1798–1800: The Early Contributors to the "Anti-Jacobin Review."* New York: St. Martin's, 1988.

Moorman, Mary, ed. *Journals of Dorothy Wordsworth*. London: Oxford UP, 1971.

Morrell, Jack, and Arnold Thackray. *Gentlemen of Science: Early Years of the British Association for the Advancement of Science*. Oxford: Clarendon, 1981.

Morton, A. G. *History of Botanical Science: An Account of the Development of Botany from Ancient Times to the Present Day*. London: Academic, 1981.

Mumm, S. D. "Writing for Their Lives: Women Applicants to the Royal Literary Fund, 1840–1880." *Publishing History* 27 (1990): 27–47.

Myers, Greg. "Fictions for Facts: The Form and Authority of the Scientific Dialogue." *History of Science* 30, 3 (1992): 221–47.

———. "Science for Women and Children: The Dialogue of Popular Science in the Nineteenth Century." In *Nature Transfigured: Science and Literature, 1700–1900*, ed. John Christie and Sally Shuttleworth. Manchester: Manchester UP, 1989.

———. *Writing Biology: Texts in the Social Construction of Scientific Knowledge*. Madison: U of Wisconsin P, 1990.

Myers, Mitzi. "Impeccable Governesses, Rational Dames, and Moral Mothers: Mary Wollstonecraft and the Female Tradition in Georgian Children's Books." *Children's Literature* 14 (1986): 31–59.

———. "Reform or Ruin: A Revolution in Female Manners." *Studies in Eighteenth-Century Culture* 11 (1982): 199–216.

Myers, Sylvia Harcstark. *The Bluestocking Circle: Women, Friendship, and the Life of the Mind in Eighteenth-Century England*. Oxford: Clarendon, 1990.

Nelmes, Ernest, and William Cuthbertson. *Curtis's Botanical Magazine Dedications, 1827–1927*. London: Quaritch, 1931.

Norwood, Vera. *Made from This Earth: American Women and Nature*. Chapel Hill: U of North Carolina P, 1993.

Nussbaum, Felicity. " 'Savage' Mothers: Narratives of Maternity in the Mid-Eighteenth Century." *Eighteenth-Century Life* 16 (1992): 163–84.

O'Brian, Patrick. *Joseph Banks: A Life*. London: Collins Harvill, 1987.

Oliver, F. W., ed. *Makers of British Botany: A Collection of Biographies by Living Botanists*. Cambridge: Cambridge UP, 1913.

Ong, Walter J. "Latin Language Study as a Renaissance Puberty Rite." *Studies in Philology* 56 (1959): 103–24.

Orr, Clarissa Campbell. "Albertine Necker de Saussure, the Mature Woman Author, and the Scientific Education of Women." *Women's Writing: The Elizabethan to Victorian Period* 2, 2 (1995).

Pascoe, Judith. "Female Botanists and the Poetry of Charlotte Smith." In *Re-visioning Romanticism: British Women Writers, 1776–1837*, ed. Carol Shiner Wilson and Joel Haefner. Philadelphia: U of Pennsylvania P, 1994.

Patterson, Elizabeth Chambers. *Mary Somerville and the Cultivation of Science, 1815–1840*. Boston: Nijhoff, 1983.

Percy, Joan. "Maria Elizabetha Jacson and her *Florist's Manual*." *Garden History* 20 (1992): 45–56.

Perl, Teri. "The Ladies' Diary or Woman's Almanack, 1704–1841." *Historia Mathematica* 6 (1979): 36–53.

Perry, Ruth. "Colonizing the Breast: Sexuality and Maternity in Eighteenth-Century England." *Journal of the History of Sexuality* 2 (1991): 204–34.

Peterson, M. Jeanne. *Family, Love, and Work in the Lives of Victorian Gentlewomen*. Bloomington: Indiana UP, 1989.

Phillips, Patricia. *The Scientific Lady: A Social History of Woman's Scientific Interests, 1520–1918*. London: Weidenfeld and Nicolson, 1990.

Pollock, Linda. *With Faith and Physic: The Life of a Tudor Gentlewoman, Lady Grace Mildmay, 1552–1620*. London: Collins and Brown, 1993.

Poovey, Mary. *The Proper Lady and the Woman Writer: Ideology as Style in the Works of Mary Wollstonecraft, Mary Shelley, and Jane Austen*. Chicago: U of Chicago P, 1984.

———. *Uneven Developments: The Ideological Work of Gender in Mid-Victorian England*. Chicago: U of Chicago P, 1988.

Porter, Dorothy, and Roy Porter. *Patient's Progress: Doctors and Doctoring in Eighteenth-Century England*. Oxford: Polity, 1989.

Porter, Roy. *English Society in the Eighteenth Century*. Rev. ed. Harmondsworth: Penguin, 1990.

———. *Health for Sale: Quackery in England, 1650–1850*. Manchester: Manchester UP, 1989.

———. "Science, Provincial Culture and Popular Opinion in Enlightenment England." *British Journal for Eighteenth-Century Studies* 3 (1980): 20–46.

Pratt, Mary Louise. *Imperial Eyes: Travel Writing and Transculturation*. London: Routledge, 1992.

Prochaska, F. K. *Women and Philanthropy in Nineteenth-Century England*. Oxford: Clarendon, 1980.

Purvis, June. *Hard Lessons: The Lives and Education of Working-Class Women in Nineteenth-Century England*. Cambridge: Polity, 1989.

Raistrick, Arthur. *Quakers in Science and Industry: Being an Account of the Quaker Contribution to Science and Industry During the Seventeenth and Eighteenth Centuries*. Newton Abbot: David and Charles, 1950.

Rauch, Alan. "A World of Faith on a Foundation of Science: Science and Religion in British Children's Literature: 1761–1878." *Children's Literature Association Quarterly* 14, 1 (1989): 13–19.

Raverat, Gwen. *Period Piece: A Cambridge Childhood*. London: Faber and Faber, 1987.

Rendall, Jane, ed. *Equal or Different: Women's Politics, 1800–1914*. Oxford: Blackwell, 1987.

Richardson, Richard. *Extracts from the Literary and Scientific Correspondence: Illustrative of the State and Progress of Botany*. Yarmouth, 1835.

Riddelsdell, H. J., ed. *Flora of Gloucestershire*. Cheltenham: Cotteswold Naturalists' Field Club, 1948.

Roscoe, S. *John Newbery and His Successors, 1740–1814: A Bibliography*. Wormley: Five Owls, 1973.

Rossiter, Margaret W. "Women in the History of Scientific Communication." *Journal of Library History* 21 (1986): 39–59.

———. *Women Scientists in America: Struggles and Strategies to 1940*. Baltimore: Johns Hopkins UP, 1982.

Rothstein, Natalie. *Silk Designs of the Eighteenth Century in the Collection of the Victoria and Albert Museum, London*. London: Thames and Hudson, 1990.

Rousseau, G. S. "Scientific Books and Their Readers in the Eighteenth Century." In *Books and Their Readers*, ed. Isobel Rivers. New York: St. Martin's, 1982.

Rudolph, Emanuel D. "How It Developed That Botany Was the Science Thought Most Suitable for Victorian Young Ladies." *Children's Literature* 2 (1973): 92–97.

Russell-Gebbett, Jean. *Henslow of Hitcham: Botanist, Educationalist and Clergyman*. Lavenham: Terence Dalton, 1977.

Russett, Cynthia Eagle. *Sexual Science: The Victorian Construction of Womanhood*. Cambridge: Harvard UP, 1989.

Sachs, Julius von. *History of Botany, 1530–1860*. Trans. H. E. F. Garnsey. 1890; rpt. New York: Russell and Russell, 1967.

Schaaf, Larry. *Out of the Shadows: Herschel, Talbot and the Invention of Photography*. New Haven: Yale UP, 1993.

———. *Sun Gardens: Victorian Photograms*. New York: Aperture, 1985.

Schiebinger, Londa. "The History and Philosophy of Women in Science: A Review Essay." *Signs* 12 (1987): 305–32.

———. *The Mind Has No Sex? Women in the Origins of Modern Science*. Cambridge: Harvard UP, 1989.

———. *Nature's Body: Gender in the Making of Modern Science*. Boston: Beacon, 1993.

Schofield, Robert E. *The Lunar Society of Birmingham: A Social History of Provincial Science and Industry in Eighteenth-Century England*. Oxford: Clarendon, 1963.

Scourse, Nicolette. *Victorians and Their Flowers*. London: Croom Helm, 1983.

Seaton, Beverly. "Considering the Lilies: Ruskin's 'Proserpina' and Other Victorian Flower Books." *Victorian Studies* 28, 2 (1985): 255–82.

Secord, Anne. "Science in the Pub: Artisan Botanists in Early Nineteenth-Century Lancashire." *History of Science* 32 (1994): 269–315.

Secord, James A. "Newton in the Nursery: Tom Telescope and the Philosophy of Tops and Balls, 1761–1838." *History of Science* 23 (1985): 127–51.

Shapin, Steven. "'A Scholar and a Gentleman': The Problematic Identity of the Scientific Practitioner in Early Modern England." *History of Science* 29 (1991): 279–328.

Shapin, Steven, and Simon Schaffer. *Leviathan and the Air-Pump: Hobbes, Boyle and the Experimental Life*. Princeton: Princeton UP, 1985.

Sheets-Pyenson, Susan. "From the North to Red Lion Court: The Creation and Early Years of the *Annals of Natural History*." *Archives of Natural History* 10, 2 (1981): 221–49.

———. "A Measure of Success: The Publication of Natural History Journals in Early Victorian Britain." *Publishing History* 9 (1981): 21–36.

———. "Popular Science Periodicals in Paris and London: The Emergence of a Low Scientific Culture, 1820–1875." *Annals of Science* 42 (1985): 549–72.

Shevelow, Kathryn. *Women and Print Culture: The Construction of Femininity in the Early Periodical*. London: Routledge, 1989.

Shiach, Morag. *Discourse on Popular Culture: Class, Gender and History in Cultural Analysis, 1730 to the Present*. Cambridge: Polity, 1989.

Shinn, Terry, and Richard Whitely, eds. *Expository Science: Forms and Functions of Popularization*. Dordrecht: Reidel, 1985.

Shteir, Ann B. "Botanical Dialogues: Maria Jacson and Women's Popular Science Writing in England." *Eighteenth-Century Studies* 23 (1990): 301–17.

———. "Botany in the Breakfast Room: Women and Early Nineteenth Century British Plant Study." In *Uneasy Careers and Intimate Lives: Women in Science, 1789–1979*, ed. Pnina G. Abir-Am and Dorinda Outram, 31–43. New Brunswick: Rutgers UP, 1987.

———. "Flora Feministica: Reflections on the Culture of Botany." *Lumen* 12 (1993): 167–76.

———. "Linnaeus's Daughters: Women and British Botany." In *Women and the Structure of Society: Selected Research from the Fifth Berkshire Conference on the History of Women*, ed. Barbara J. Harris and Joann K. McNamara, 67–73. Durham, N.C.: Duke UP, 1984.

———. "Priscilla Wakefield's Natural History Books." In *From Linnaeus to Darwin: Commentaries on the History of Biology and Geology*, 29–36. London: Society for the History of Natural History, 1985.

Simmons, Mark. *A Catalogue of the Herbarium of the British Flora Collected by Margaret Stovin (1756–1846)*. Middlesbrough: Dorman Museum, 1993.

Smith, Olivia. *The Politics of Language, 1791–1819*. Oxford: Clarendon, 1984.

Society of Friends. *Annual Monitor*. 1856.

Sowerby, Arthur DeCarle. *The Sowerby Saga*. Washington, D.C., 1952.

Spencer, Jane. *The Rise of the Woman Novelist: From Aphra Behn to Jane Austen*. Oxford: Blackwell, 1982.

Stafford, Barbara Maria. *Body Criticism: Imaging the Unseen in Enlightenment Art and Medicine*. Cambridge: MIT P, 1991.

Stafleu, Frans A. *Linnaeus and the Linnaeans: The Spreading of Their Ideas in Systematic Botany, 1735–1789*. Utrecht: International Association for Plant Taxonomy, 1971.

Stanton, Judith Phillips. "Charlotte Smith's 'Literary Business': Income, Patronage, and Indigence." In *The Age of Johnson: A Scholarly Annual*, ed. Paul Korshin. New York: AMS, 1987.

———. "Statistical Profile of Women Writing in English from 1660 to 1800." In *Eighteenth-Century Women and the Arts*, ed. Frederick M. Keener and Susan E. Lorsch. New York: Greenwood, 1988.

Stevens, Peter F. *The Development of Biological Systematics: Antoine-Laurent de Jussieu, Nature and the Natural System*. New York: Columbia UP, 1994.

Stewart, Larry. *The Rise of Public Science: Rhetoric, Technology, and Natural Philosophy in Newtonian Britain, 1660–1750*. New York: Cambridge UP, 1992.

Sutherland, John. "Henry Colburn, Publisher." *Publishing History* 19 (1986): 59–84.

Tatchell, Molly. "Elizabeth Kent and *Flora Domestica*." *Keats-Shelley Memorial Bulletin* 27 (1976): 15–18.

———. *Leigh Hunt and His Family in Hammersmith*. London: Hammersmith Local History Group, 1969.

Taylor, Geoffrey. *Some Nineteenth Century Gardeners*. London, 1951.

Teiman, Gillian. "The Female Ideal and the Female Voice: Ideology, Resistance and Accommodation in the *Tatler*, and *Spectator*, the *Female Tatler*, and the *Female Spectator*." Ph.D. diss., York University, 1992.

Thaddeus, Janice Farrar. "Mary Delany, Model to the Age." In *History, Gender and Eighteenth-Century Literature*, ed. Beth Fowkes Tobin. Athens: U Georgia P, 1994.

Thomas, Keith. *Man and the Natural World: A History of the Modern Sensibility*. New York: Pantheon, 1983.

Todd, Janet, ed. *A Dictionary of British and American Women Writers, 1660–1800*. Totowa, N.J.: Rowman and Allenheld, 1985.

———. *The Sign of Angellica: Women, Writing and Fiction, 1660–1800*. London: Virago, 1989.

Topham, Jonathan. "Science and Popular Education in the 1830s: The Role of the *Bridgewater Treatises*." *British Journal for the History of Science* 25 (1992): 397–430.

Traill, T. S. *Memoir of Wm. Roscoe*. Liverpool, 1853.

Tuana, Nancy, ed. *Feminism and Science*. Bloomington: Indiana UP, 1989.

Tuchman, Gaye, with Nina E. Fortin. *Edging Women Out: Victorian Novelists, Publishers, and Social Change*. New Haven: Yale UP, 1989.

Turner, Frank M. *Contesting Cultural Authority: Essays in Victorian Intellectual Life*. Cambridge: Cambridge UP, 1993.

Walker, Margot. "Admission of Lady Fellows." *The Linnean* 1, 1 (1984): 9–11.

———. *Sir James Edward Smith M.D., F.R.S., P.L.S., 1759–1828: First President of the Linnean Society of London*. London: Linnean Society, 1988.

Warren, Leland. "Turning Reality Round Together: Guides to Conversation in Eighteenth-Century England." *Eighteenth-Century Life*, n.s., 8 (May 1983): 65–87.

Waters, Michael. *The Garden in Victorian Literature*. Aldershot: Scolar, 1988.

Watson, Nicola J. *Revolution and the Form of the British Novel, 1790–1825: Intercepted Letters, Interrupted Seductions*. Oxford: Clarendon, 1994.

Webb, Rev. William. *Memorials of Exmouth*. 1872.

Willson, E. J. *James Lee and the Vineyard Nursery, Hammersmith*. London: Hammersmith Local History Group, 1961.

Wolff, Janet, and John Seed, eds. *The Culture of Capital: Art, Power and the Nineteenth-Century Middle Class*. Manchester: Manchester UP, 1988.

Wystrach, V. P. "Anna Blackburne (1726–1793) — a Neglected Patron of Natural History." *Journal of the Society for the Bibliography of Natural History* 8 (1977): 148–68.

Yeldham, Charlotte. *Women Artists in Nineteenth-Century France and England.* New York: Garland, 1984.

Yeo, Richard, "Science and the Organization of Knowledge in British Dictionaries of Arts and Sciences, 1730–1850." *Isis* 82, 311 (1991): 43–48.

Young, Robert M. *Darwin's Metaphor: Nature's Place in Victorian Culture.* Cambridge: Cambridge UP, 1985.

Index

Page references to illustrations are printed in italic type.

ANN B. SHTEIR is associate professor of humanities and director of the Graduate Programme in Women's Studies, York University, Canada. She was educated at Douglass College, at Ludwig-Maximilians-Universität in Munich, and at Oxford University; she received her Ph.D. in comparative literature from Rutgers University. Her articles have appeared in *Eighteenth-Century Studies* and other professional journals. She has contributed multiple entries to *The Dictionary of British and American Women Writers, 1660–1800* (Rowman and Allanheld, 1984) and *The Feminist Companion to Literature in English* (Yale, 1990). She edited Priscilla Wakefield's popular miscellany from 1794, *Mental Improvement, or The Beauties and Wonders of Nature and Art* (Colleagues Press, 1995). She is guest editor of a special issue on "Women and Science" for the journal *Women's Writing: The Elizabethan to Victorian Period*, vol. 2, no. 2 (1995). She is coeditor with Barbara T. Gates of the forthcoming collection of essays on women and science popularization, *Science in the Vernacular*.